T0400931

ENVIRONMENTAL SCIENCE, ENGINEERING AND TECHNOLOGY

VOLATILE ORGANIC COMPOUNDS

OCCURRENCE, BEHAVIOR AND ECOLOGICAL IMPLICATIONS

ENVIRONMENTAL SCIENCE, ENGINEERING AND TECHNOLOGY

Additional books in this series can be found on Nova's website under the Series tab.

Additional e-books in this series can be found on Nova's website under the e-books tab.

ENVIRONMENTAL SCIENCE, ENGINEERING AND TECHNOLOGY

VOLATILE ORGANIC COMPOUNDS

OCCURRENCE, BEHAVIOR AND ECOLOGICAL IMPLICATIONS

JULIAN PATRICK MOORE
EDITOR

New York

Library of Congress Cataloging-in-Publication Data

Names: Moore, Julian Patrick, editor.
Title: Volatile organic compounds : occurrence, behavior and ecological implications / editor, Julian Patrick Moore.
Other titles: Volatile organic compounds (Moore)
Description: Hauppauge, New York : Nova Science Publishers Inc., [2016] |
Series: Environmental science, engineering and technology |
Includes bibliographical references and index.
Identifiers: LCCN 2016021226 (print) | LCCN 2016032481 (ebook) | ISBN 9781634853705 (hardcover) | ISBN 9781634853903 (Ebook) | ISBN 9781634853903 ()
Subjects: LCSH: Volatile organic compounds.
Classification: LCC TD196.O73 V54 2016 (print) | LCC TD196.O73 (ebook) | DDC 628.5/3--dc23
LC record available at https://lccn.loc.gov/2016021226

Published by Nova Science Publishers, Inc. † New York

CONTENTS

PREFACE

Volatile organic compounds (VOCs) are hazardous highly toxic pollutants that cause a number of environmental and human health problems. They are released during a wide range of industrial, transportation and commercial activities and their emissions have reached high levels. This book provides a review of the occurrence, behavior and ecological implications of VOCs. Chapter One evaluates the biogeneration of volatile organic compounds produced by microalgae. Chapter Two investigates VOC pollution from industrial complexes and wastewater treatment plants (WWTPs). Chapter Three illustrates the current knowledge of intra- and inter- organismal Microbial Volatile Organic Compounds (mVOCs)-based interactions, volatile perception, signal transduction and phenotypical responses in the receiver organisms. Chapter Four summarizes the current state of knowledge regarding the impact of emissions from gasoline powered vehicles on historical ambient VOC concentration trends, and on the current ambient urban atmosphere of the South Coast Air Basin. Chapter Five discusses the oxidation of VOCs over cyptomelane catalysts. Chapter Six studies VOCs from truffles and false truffles from Basilicata in Southern Italy. Chapter Seven discusses the removal of VOCs in the air by total catalytic oxidation promoted by catalysts.

Chapter 1 – Microalgae are a source of potential commercial interest biomolecules due to their diverse metabolic profile, able to synthesize different classes of organic compounds. The continual growth of the commercial application of primary and secondary biotechnology metabolites and more strict environmental legislations have led to interest in developing renewable forms to produce these compounds and apply in bulk and fine chemistry. The growing interest in natural products directs the development of technologies that employ microorganisms, including microalgae, which are able to synthesize specific volatile organic compounds (VOCs). The different VOCs can belong to different classes of compounds such as alcohols, esters, hydrocarbons, terpenes, ketones, carboxylic acids and sulfurized compounds. Volatile organic compounds are secondary metabolites obtained from microalgae that could be used as an important alternative source of chemicals. The use of the volatile fraction of microalgal systems may represent an improvement in the supply of a large volume of inputs to many different types of industry. Clearly, there is a need for further studies on the volatile fraction of microalgal systems, as well as on the elucidation of the formation metabolic pathways of these compounds. Exploring the volatile profile of microalgae is a possibility, and it is scientifically challenging to apply these metabolites as chemical feedstocks. Divided into three discrete parts, the chapter covers topics that refer to the occurrence and behavior of volatile organic compounds in microalgae systems, the ecological implications and industrial applications, summarizing a range of useful technological and economic opportunities regarding such compounds.

Chapter 2 – The emission of volatile organic compounds (VOCs) from industrial stationary sources is becoming an environmental issue of increasing concern. Industrial complexes and wastewater treatment plants (WWTPs) represent two important VOC sources from industrial activites. This chapter attempted to investigate the VOC pollutions from industrial complexes and WWTPs by starting from summarizing selected articles and discussing the effects of altitude and tropical savanna climate on the VOC distributions in Kaohsiung City of southern Taiwan. Furthermore, additional articles that investigated the VOC emission rates, mass distirbutions among different phases (e.g., air, water, and sludge), and the associated adverse health risks to local workers and populations living near a WWTP were selected for discussion. In the first discussion, the the altitude effect on VOC distributions was introduced by analyzing the VOC concentrations at ground level and three different altitudes and assessing the results by principal component analysis (PCA) and cluster analysis. Next, the influences of diurnal temperature and seasonal humidity variations by tropical savanna climate on the distributions of VOCs were investigated. For the VOC emissions in WWTPs, biological treamtent is the process more likely to be relevant to the VOC productions and modified to mitigate the VOC emissions. By conducting lab-scale batch simulation experiments, this capter futher discussed the fates of aromatic and chlorinated hydrocarbons in wastewater treatment processes, with respect to the difference amongst the species, together with the effects of aeration and the presence of activated sludge. With the growing concern regarding the VOC emissions from WWTPs, the relationship between the VOC emission rates and the associated public health risks was also compared and examined by using a municipal WWTP in northern China as an example. Last, the VOC distributions through the full-scale WWTP were fully investigated with respect to the effects of seasonal temperature and treatment technology variations, followed by the adverse health risk assessment that simulated the atmospheric behaviors of the VOCs emitted from the WWTP and calculated the associated cancer and non-cancer risks. The discussion here followed that the VOCs emitted from a WWTP could be one important concern for the chronic adverse health risks for both the workers in the WWTP and those people living some distances away from the plant. It emphasized the complex nature of VOC emissions from WWTPs and quantitatively indicated that the associated health impacts to the public near the WWTPs could be severely underestimated, whereas their treatment efficiencies by wastewater treatment technologies were overestimated.

Chapter 3 – Volatiles are ubiquitously present on Earth due to their physical and chemical properties. Whereas animal and plant volatile emissions have been comprehensively studied in the past, volatiles of microorganisms have been neglected. The wealth of Microbial Volatile Organic Compounds (mVOCs) has been recently discovered. Besides the elucidation of their chemical structures, unraveling the biological functions of mVOCs will be one of the major tasks in the future. Microbial VOCs play important biological roles in multitrophic interactions. The present chapter will illustrate the current knowledge of intra- and inter- organismal mVOC-based interactions, volatile perception, signal transduction and phenotypical responses in the receiver organisms.

Chapter 4 – Volatile Organic Compounds (VOCs) play a significant role in the chemistry of air pollution at the local, regional, and global level by contributing to the formation of ozone and secondary organic aerosols (SOA). Some VOCs, including benzene, 1,3-butadiene, formaldehyde, and acetaldehyde are carcinogenic. Emissions from gasoline powered vehicles were known to be the main contributor to ambient VOC levels in all major cities of the world including the South Coast Air Basin (SoCAB) beginning in the 1960s. During the last five decades California led the world in taking action to reduce gasoline vehicle emissions.

Ambient VOC monitoring data have shown a significant decrease (>90%) in ambient VOC concentrations in the SoCAB since the 1960s. Studies using data from ambient monitoring, highway tunnels, roadside measurements, and chassis dynamometers testing, showed that gasoline vehicle emission control required by regulatory agencies reduced emissions from gasoline powered vehicles more than 99 percent since the 1970s. This reduction of gasoline vehicle emissions is the main factor contributing to the decreases in ambient VOC concentrations in the SoCAB.

Given the large emission reductions from motor vehicles, it was anticipated that ambient VOCs would be dominated by sources other than tailpipe emissions and current emission inventories reflect this assessment. However, several recent inventory evaluation studies, based on the analysis of observed ambient data, showed that current emission inventories significantly underestimated gasoline powered vehicle emissions. One tunnel study indicated that the evaporative emissions of gasoline powered vehicles on hot days might be the major contributor to current ambient VOCs. However a recent study comparing trends in ambient VOC concentrations to trends in VOC emissions from gasoline vehicles tested on chassis dynamometers and to trends observed in tunnel studies showed that tailpipe emissions remained the main contributor to ambient VOCs in the SoCAB, in contradiction to the current inventory estimate and to studies suggesting that evaporative emissions explain the discrepancy. Control of tailpipe emissions from gasoline vehicles continues to be the best strategy for reducing ambient VOC levels and their harmful reaction products.

Chapter 5 – Volatile organic compounds (VOC) are hazardous environmental pollutants originated from different sources, such as petroleum refineries, fuel storage, motor vehicles, painting and printing activities. VOCs emissions are regulated due to their potential damages to human health and the environment. Catalytic oxidation to CO_2 and H_2O is an environmentally friendly technology for VOC abatement that needs low temperatures (around 250-500ºC) and causes less NOx formation, compared to conventional thermal oxidation, which operates at higher temperatures (650-1100ºC). Several catalysts have been used for this purpose.

In this work, cryptomelane-type manganese oxides were tested for the oxidation of ethyl acetate and toluene, two common VOCs. Catalysts were synthesized by redox reaction under acid and reflux conditions. Different metals (cerium, cesium and lithium) were incorporated into the tunnel structure of cryptomelane by the ion-exchange technique. Gold was loaded onto these materials (1% wt.) by a double impregnation method. The obtained catalysts were characterized by X-ray diffraction, high-resolution transmission electron microscopy, energy-dispersive X-ray diffraction and temperature programmed reduction. It was found that addition of Cs and Li to cryptomelane was beneficial for ethyl acetate oxidation, but addition of Ce is detrimental, which is related to the reducibility of the materials. Addition of gold does not improve the catalytic activity, and in case of Li and Cs modified samples, it has even an unfavourable effect. This effect can be related to the large gold particle size found for these samples, well known to be inversely related with catalytic activity. Addition of Li to cryptomelane was beneficial for toluene oxidation, but adding Cs and Ce was disadvantageous. Loading with gold also did not show considerable improvement, which may also be related to particle size effects. The activity for both VOCs abatement was correlated with sample reducibility.

Chapter 6 – Volatile organic compounds (VOCs) of several Tuber species are identified via solid-phase microextraction-gas-chromatography-mass spectrometry analysis. The VOCs of *T. mesentericum*, *T. exacavatum*, *T. borchii*, *T. magnatum*, *T. aestivm*, *T. uncinatum*, *T. brumale*, *T. melanosporum*, *T. oligospermum*, *T. panniferum* have been determined.

Ascomata of two truffle species, *Tuber borchii* and *T. asa-foetida* were identified on the base of ascospore morphology and compared under volatile organic compound profile to determine the particular VOCs which characterize each taxon. SPME-GC-MS analysis of the samples showed the presence of 1-methyl-1,3-butadiene as a main component in both the truffles. *T. asa-foetida* showed a compound, toluene, not present in *T. borchii*, in agreement with a penetrant "solvent" smell of the truffle.

Volatile organic compounds (VOCs) of *Schenella pityophilus* have been identified via solid-phase microextraction-gas chromatography-mass spectrometry analysis. Ten compounds have been identified. 3-Methylthio-1-propene was the most significant component. Some other components were identified previously in *Tuber aestivum* and *Tuber melanosporum*.

Results of SPME-CG-MS analyses, accomplished on sporophores of eleven species of truffles and false truffles, are reported. VOCs found in *Gautieria morchelliformis* were dimethyl sulphide, 1,3-octadiene, 3,7-dimethyl-1,6-octadien-3-ol, γ-muurolene, amorphadiene, isoledene, and *cis*-muurola-3,5-diene. In *Hymenogaster luteus* var. *luteus,* presence of 1,3-octadiene, 1-octen-3-ol, 3-octanone, 3-octanol, and 4-acetylanisole was revealed. Two VOCs, 4-acetylanisole and β-farnesene, constituted aroma of *Hymenogaster olivaceus. Melanogaster broomeanus* exhibited as components of its aroma 2-methyl-1,3-butadiene, 2-methylpropanal, 2-methylpropanol, isobutyl acetate, 3,7-dimethyl-1,6-octadien-3-ol, 3-octanone, and β-curcumene. *Melanogaster variegatus* showed the presence of 2-methylpropanol, ethyl 2-methylpropanoate, isobutyl acetale, 2-methypropyl 2-methyl-2-butenoate, 3-phenylpropyl acetate, 2-methylpropyl propanoate. VOC profile of *Octavianina asterosperma* was characterized by the presence of dimethyl sulphide, ethyl 2-methylpropanoate, methyl 2-methylbutanoate, and 3-octanone. *Choiromyces meandriformis*, *Tuber rufum* var. *rufum*, *T. rufum* var. *lucidum* and *Pachyphloeus conglomeratus* showed the presence of dimethyl sulphide only. Some methoxy substitued ethers were, finally, present in *Tuber dryophilum*.

Chapter 7 – Volatile organic compounds (VOCs) are hazardous highly toxic pollutants that cause a number of environmental and human health problems. They are released during a wide range of industrial, transportation and commercial activities and their emissions have reached high levels. There is therefore a need for the development of techniques which are both economically feasible and able to effectively remove these pollutants. Several VOCs removal techniques, including physical, chemical and biological methods, have been described in the literature. Catalytic oxidation has been acknowledged as the most effective approach, mainly due to its high degradation efficiency, low energy cost, the potential for the removal of low concentrations of VOCs and the low thermal NO_x emissions involved. This capther was to investigate the catalytic behavior of catalysts with low production costs, involving a simple method of preparation without the use of reagents which are harmful to the environment, which are active at low temperature, have high redox potential for catalytic oxidation reactions and do not result in the formation of byproducts. In this regard, catalysts $Co_{10}/\gamma\text{-}Al_2O_3\text{-}CeO_2$ and $Co_{20}/\gamma\text{-}Al_2O_3\text{-}CeO_2$; catalysts ($SiO_{2(1\text{-}}x)Cux$) were developed which exhibit high catalytic activity for converting BTX compounds and WO_3-based catalyst to investigate its activity catalytic in the total oxidation of the volatile organic compounds known as BTX. For a range of low temperatures the only reaction products were CO_2 and H_2O.

In: Volatile Organic Compounds
Editor: Julian Patrick Moore
ISBN: 978-1-63485-370-5

Chapter 1

BIOGENERATION OF VOLATILE ORGANIC COMPOUNDS BY MICROALGAE: OCCURRENCE, BEHAVIOR, ECOLOGICAL IMPLICATIONS AND INDUSTRIAL APPLICATIONS

Andriéli B. Santos, Karem R. Vieira, Gabriela P. Nogara, Roger Wagner, Eduardo Jacob-Lopes and Leila Q. Zepka
Department of Food Science and Technology,
Federal University of Santa Maria (UFSM), Santa Maria, RS, Brazil

ABSTRACT

Microalgae are a source of potential commercial interest biomolecules due to their diverse metabolic profile, able to synthesize different classes of organic compounds. The continual growth of the commercial application of primary and secondary biotechnology metabolites and more strict environmental legislations have led to interest in developing renewable forms to produce these compounds and apply in bulk and fine chemistry. The growing interest in natural products directs the development of technologies that employ microorganisms, including microalgae, which are able to synthesize specific volatile organic compounds (VOCs). The different VOCs can belong to different classes of compounds such as alcohols, esters, hydrocarbons, terpenes, ketones, carboxylic acids and sulfurized compounds. Volatile organic compounds are secondary metabolites obtained from microalgae that could be used as an important alternative source of chemicals. The use of the volatile fraction of microalgal systems may represent an improvement in the supply of a large volume of inputs to many different types of industry. Clearly, there is a need for further studies on the volatile fraction of microalgal systems, as well as on the elucidation of the formation metabolic pathways of these compounds. Exploring the volatile profile of microalgae is a possibility, and it is scientifically challenging to apply these metabolites as chemical feedstocks. Divided into three discrete parts, the chapter covers topics that refer to the occurrence and behavior of volatile organic compounds in microalgae systems, the ecological implications and industrial applications, summarizing a range of useful technological and economic opportunities regarding such compounds.

Keywords: biomolecule, biossynthesis, flavor, off-flavor, fuel

INTRODUCTION

Microalgae are a group of photosynthetic microorganisms typically unicellular and eukaryotic. Although cyanobacteria belong to the domain of bacteria, and are photosynthetic prokaryotes, often they are considered microalgae [1].

Microalgae-based systems for chemicals production are an emergent area, representing a great promise for industrial application. However, there is little information available on the volatile organic compounds biogeneration of these microorganisms. The characterization of the volatile fraction of microalgal bioreactors can contribute to establishing routes for the bioconversion of substrates, and enable the identification of potential applications of the volatile bioproducts formed [2].

The growing interest in natural products guides the development of technologies that employ microorganisms, including microalgae, which are able to synthesize specific volatile organic compounds (VOCs). Jacob-Lopes [3] reported that the VOCs are the main bioproducts formed during microalgae cultivation. The carbon balance analysis indicates that these compounds represent up to 90% of the total substrate converted in the bioreactor. The different VOCs can belong to different classes of compounds such as alcohol, esters, hydrocarbons, terpenes, ketones, carboxylic acids and sulfurized compounds [4].

Microalgae were always regarded to be typical photosynthetic microorganism in which the light-dependent fixation of CO_2 is the dominant mode of nutrition [5]. Microalgae can also be cultivated heterotrophically without light and with addition of an exogenous source of carbon by using the oxidative pentose phosphate pathway. This metabolic route serves as the exclusive source of energy for maintenance and biosynthesis, besides providing the carbon required as building blocks for biosynthesis [6]. The biosynthesis of volatile compounds depends mainly of the availability of carbon and nitrogen as well as energy provided by primary metabolism. Therefore, the availability of these building blocks has a major impact on the concentration of any secondary metabolites, including VOCs [2].

Based on their biosynthetic origin, these VOCs can be divided into terpenoids, phenylpropanoids/benzenoids, carbohydrate derivates, fatty acids derivates and amino acid derivates, in addition to specific compounds not represented in those major classes [7; 8]. These compounds could therefore be a source of useful chemical products, based on a nonconventional technological route. Chemicals obtained from bioprocesses are sold at prices 1000 times higher than those synthetic chemicals, which show great potential for the exploitation of these processes. In view of the commercial significance, efforts should be made to elucidate the pathways of the formation of these compounds. Thus, the aim of this chapter was to evaluate the biogeneration of volatile organic compounds produced by microalgae, with focus on the occurrence, behavior, ecological implications and industrial applications of these metabolites.

OCCURRENCE AND BEHAVIOROF VOLATILE ORGANIC COMPOUNDS FROM MICROALGAE

The occurrence of volatile organic compounds in microalgae is a consequence of their metabolism. Microalga species use oxygenic photosynthesis for the fixation of CO_2 [9]. They have pigments such as chlorophyll and carotenoids, which are involved in capturing luminous energy to perform photosynthesis. For the CO_2 be converted into carbohydrates, catalyzed by the enzyme ribulose 1,5-bisphosphate carboxylase/oxygenase (Rubisco), is referred to as the Calvin cycle. The Calvin cycle is the metabolic mechanism for fixing CO_2 in microalgae. This process comprises three stages; carboxylation, reduction and regeneration [10]. The end of the cycle to form one molecule of glyceraldehyde-3-phosphate that through the action of enzymes form phosphoenolpyruvate, and finally pyruvate.

Additionally, some species of these microorganisms have the versatility to maintain their structures in the absence of light, being able to grow heterotrophically through the assimilation of one or more organic substrates as a carbon source in the oxidative pentose phosphate pathway. To use these organic compounds, transport occurs through the membrane. This substrate will be converted into glucose 6-phosphate so you can start the route. During metabolism there is the formation of two molecules of ATP (adenosine triphosphate). The final product is also pyruvate [5].

Regardless of metabolism, the biosynthesis of volatile organic compounds occurs through the formation of pyruvate molecule. To further illustrate, Figure 1 shows the main pathways of formation of these compounds which may be enzymatically or by a reaction degradation. Based on this knowledge, can be suggested applicable routes for the synthesis of volatile organic compounds both for their better ecological understanding as to their potential commercial applications.

The volatile organic compounds from microalgae can belong to different classes of compounds such as esters, alcohols, hydrocarbons, ketones, terpenes, carboxylic acids and sulfur compounds [11]. In order to understand this diversity of compounds, Table 1 shows the main compounds and their cultivations previously found in studies. The biosynthesis of these volatile organic compounds will depend on the availability of building blocks, such as carbon, nitrogen, and energy supply from the primary metabolism. Therefore, the availability of these building blocks has great impact on the concentration of secondary metabolites, including VOCs, demonstrating the high level of connectivity between the primary and secondary metabolism [12, 13].

The formation of compounds from pyruvate can follow the route of terpenoids or also via the keto acids via intermediate 2-ketoisovalerate. With the formation of Acetyl-CoA has the biosynthesis of fatty acids. On arriving at the tricarboxylic acid cycle, follows the route of keto acids by intermediates 2-ketobutyrate and 2-ketovalerate. Started by way 2-keto acids, amino acids, which are intermediates in the synthesis pathways. These 2-keto acids are formed by deamination followed by decarboxylation catalyzed by transaminase branched chain amino acids such as L-leucine, which has its synthesis from pyruvate, and L-isoleucine from the tricarboxylic acid or also by intermediate amino acid synthesis [14]. These 2-keto acids can be further subjected to decarboxylation, followed by reduction, oxidation and/or esterification, can be formed in addition to alcohols, aldehydes, acids and esters [12].

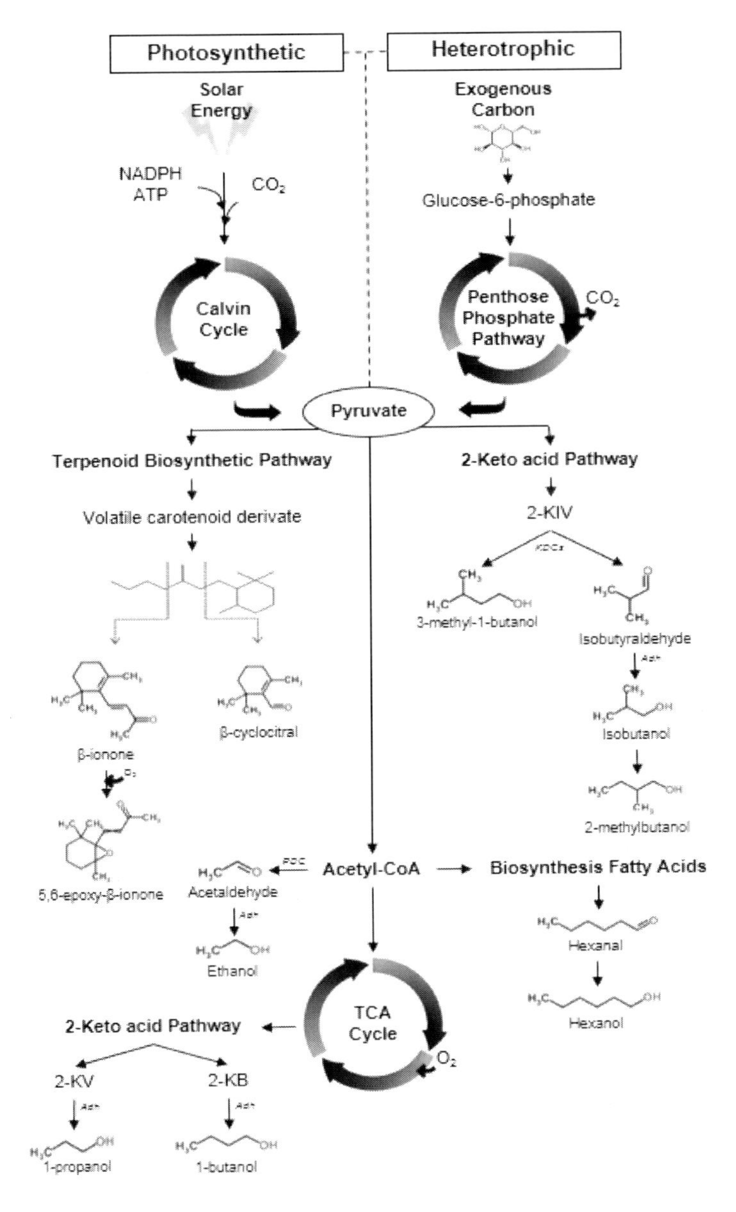

Figure 1. Overview of microalgae metabolism and their potential routes emission of volatile organic compounds.

Table 1. Major VOCs found in microalgae-based systems

Organic classes	Compounds	Microalgae	Cultivation	Reference
Acids	Acetic acid	*Rhodomonas* sp.; *Nannochloropsis oculata; Botryococcus braunii; Chlorella vulgaris*	Photoautotrofic	Durme et al., 2013 [8]
		Phormidium autumnale	Heterotrophic	Santos et al., 2015 [13]
	Butanoic acid	*Phormidium autumnale*	Heterotrophic	Santos et al., 2015 [13]
	3-Methylbutanoic acid	*Phormidium autumnale*	Heterotrophic	Santos et al., 2015 [13]
	Isovaleric acid	*Phormidium autumnale*	Heterotrophic	Santos et al., 2015 [13]
Sulfuric compounds	Dimethyl sulfide	*Tetraselmis* sp.	Photoautotrofic	Durme et al., 2013 [8]
	Dimethyl disulfide	*Tetraselmis* sp.; *Rhodomonas* sp.	Photoautotrofic	Durme et al., 2013 [8]
	Methional	*Rhodomonas* sp.	Photoautotrofic	Durme et al., 2013 [8]
	Dimethyl trisulfide	*Tetraselmis* sp.	Photoautotrofic	Durme et al., 2013 [8]
Furans	2-Ethylfuran	*Botryococcus braunii*	Photoautotrofic	Durme et al., 2013 [8]
	2-Pentylfuran	*Botryococcus braunii; Chlorella vulgaris; Spirulina platensis*	Photoautotrofic	Durme et al., 2013 [8]
Esters	Ethyl acetate	*Nannochloropsis oculata*	Photoautotrofic	Durme et al., 2013 [8]
		Phormidium autumnale	Heterotrophic	Santos et al., 2015 [13]
	Methyl hexanoate	*Nannochloropsis oculata; Chlorella vulgaris*	Photoautotrofic	Durme et al., 2013 [8]
	Methyl phenylacetate	*Nannochloropsis oculata*	Photoautotrofic	Durme et al., 2013 [8]
	Methyl octanoate	*Nannochloropsis oculata; Rhodomonas* sp.; *Chlorella vulgaris; Botryococcus braunii*	Photoautotrofic	Durme et al., 2013 [8]
	Isoamyl acetate	*Phormidium autumnale*	Heterotrophic	Santos et al., 2015 [13]
	Isobutyl acetate	*Phormidium autumnale*	Heterotrophic	Santos et al., 2015 [13]

Organic classes	Compounds	Microalgae	Cultivation	Reference
	Methyl decanoate	*Nannochloropsis oculata*	Photoautotrofic	Durme et al., 2013 [8]
	2-Butyl acetate	*Phormidium autumnale*	Heterotrophic	Santos et al., 2015 [13]
Terpenes	β-Cyclocitral	*Tetraselmis* sp.; *Rhodomonas* sp.; *Nannochloropsis oculata; Botryococcus braunii; Chlorella vulgaris; Spirulina platensis; Nostoc* sp.; *Anabaena* sp.	Photoautotrofic	Durme et al., 2013 [8]; Milovanovic et al., 2015 [38]
		Phormidium autumnale;	Heterotrophic	Santos et al., 2015 [13]
	β-Ionone	*Tetraselmis* sp.: *Rhodomonas* sp.; *Chlorella vulgaris;*	Photoautotrofic	Durme et al., 2013 [8]
		Phormidium autumnale	Heterotrophic	Santos et al., 2015 [13]
Ketones	2,3-Butanedione	*Tetraselmis* sp.; *Rhodomonas* sp.; *Nannochloropsis oculata; Chlorella vulgaris*	Photoautotrofic	Durme et al., 2013 [8]
		Phormidium autumnale	Heterotrophic	Santos et al., 2015 [13]
	2-Butanone	*Phormidium autumnale*	Heterotrophic	Santos et al., 2015 [13]
	3-Methyl-2-butanone	*Phormidium autumnale*	Heterotrophic	Santos et al., 2015 [13]
	4-Methyl-2-pentanone	*Phormidium autumnale*	Heterotrophic	Santos et al., 2015 [13]
	1-Penten-3-one	*Tetraselmis* sp.; *Rhodomonas* sp.; *Nannochloropsis oculata; Botryococcus braunii; Chlorella vulgaris*	Photoautotrofic	Durme et al., 2013 [8]
	2,3-Pentanedione	*Tetraselmis* sp.; *Rhodomonas* sp.	Photoautotrofic	Durme et al., 2013 [8]
	Acetophenone	*Phormidium autumnale*	Heterotrophic	Santos et al., 2015 [13]
	2,3-hexanedione	*Phormidium autumnale*	Heterotrophic	Santos et al.,2015 [13]
	3-Pentanone	*Nannochloropsis oculata ; Botryococcus braunii; Chlorella vulgaris*	Photoautotrofic	Durme et al., 2013 [8]
	2-Heptanone	*Phormidium autumnale*	Heterotrophic	Santos et al., 2015 [13]

Table 1. (Continued)

Organic classes	Compounds	Microalgae	Cultivation	Reference
	3-Hydroxy-2-butanone	*Rhodomonas* sp.; *Nannochloropsis oculata; Chlorella vulgaris*	Photoautotrofic	Durme et al., 2013 [8]
	2,3-Octanedione	*Rhodomonas* sp.;	Photoautotrofic	Durme et al., 2013 [8]
	6-Methyl-5-hepten-2-one	*Tetraselmis* sp.;	Photoautotrofic	Durme et al., 2013 [8]
	Tr,tr-3,5-octadien-2-one	*Rhodomonas*	Photoautotrofic	Durme et al., 2013 [8]
	6-Methyl-2-heptanone	*Phormidium autumnale*	Heterotrophic	Santos et al., 2015 [13]
Alcohols	Ethanol	*Tetraselmis* sp.; *Chlorella vulgaris; Nannochloropsis oculata*	Photoautotrofic	Durme et al., 2013 [8]
	1-Penten-3-ol	*Tetraselmis* sp.; *Rhodomonas* sp.; *Nannochloropsis oculata; Chlorella vulgaris; Botryococcus braunii*	Photoautotrofic	Durme et al., 2013 [8]
	3-Methylbutanol	*Tetraselmis* sp.; *Nannochloropsis oculata; Chlorella vulgaris*	Photoautotrofic	Durme et al., 2013 [8]
		Phormidium autumnale	Heterotrophic	Santos et al., 2015 [13]
	2-Methylbutanol	*Tetraselmis* sp.; *Nannochloropsis oculata; Chlorella vulgaris*	Photoautotrofic	Durme et al., 2013 [8]
	2-Methyl-1-propanol	*Phormidium autumnale*	Heterotrophic	Santos et al., 2015 [13]
	2-Phenylethanol	*Phormidium autumnale*	Heterotrophic	Santos et al., 2015 [13]
	1-Pentanol	*Tetraselmis* sp.; *Rhodomonas* sp.; *Nannochloropsis oculata; Chlorella vulgaris; Botryococcus braunii*	Photoautotrofic	Durme et al., 2013 [8]
	Cis-2-penten-1-ol	*Tetraselmis* sp.; *Rhodomonas* sp.; *Botryococcus; Chlorella vulgaris*	Photoautotrofic	Durme et al., 2013 [8]

Organic classes	Compounds	Microalgae	Cultivation	Reference
	3-Hexen-1-ol	*Chlorella vulgaris*	Photoautotrofic	Durme et al., 2013 [8]
	1-Hexanol	*Tetraselmis* sp.; *Nannochloropsis oculata; Chlorella vulgaris*	Photoautotrofic	Durme et al., 2013 [8]; Santos et al., 2015 [13]
		Phormidium autumnale	Heterotrophic	Santos et al., 2015 [13]
	2,6-Dimethylcyclohexanol	*Spirulina platensis; Nostoc* sp.; *Anabaena* sp.	Photoautotrofic	Milovanovic et al., 2015 [38]
	2-Ethyl-1-hexanol	*Spirulina platensis; Nostoc* sp.	Photoautotrofic	Milovanovic et al., 2015 [38]
	1-Octen-3-ol	*Rhodomonas; Nannochloropsis oculata;*	Photoautotrofic	Durme et al., 2013 [8]
	2-Pentanol	*Phormidium autumnale*	Heterotrophic	Santos et al., 2015 [13]
	2-Propanol	*Phormidium autumnale*	Heterotrophic	Santos et al., 2015 [13]
	Butanol	*Phormidium autumnale*	Heterotrophic	Santos et al., 2015 [13]
	3-Methyl-1-butanol	*Phormidium autumnale*	Heterotrophic	Santos et al., 2015 [13]
	Nonanol	*Phormidium autumnale*	Heterotrophic	Santos et al, 2015 [13]
	2-Ethylhexanol	*Phormidium autumnale*	Heterotrophic	Santos et al., 2015 [13]
		Spirulina platensis; Nostoc sp.	Photoautotrofic	Milovanovic et al., 2015 [38]
	1-heptanol	*Phormidium autumnale*	Heterotrophic	Santos et al., 2015 [13]
	3-Methylbutanoic acid	*Phormidium autumnale*	Heterotrophic	Santos et al., 2015 [13]
Aldehydes	Pentanal	*Rhodomonas* sp.; *Botryococcus braunii*	Photoautotrofic	Durme et al., 2013 [8]
	Ethyl-3-hydroxybutanoate	*Phormidium autumnale*	Heterotrophic	Santo et al., 2015 [13]
	Decanal	*Phormidium autumnale*	Heterotrophic	Santos et al., 2015 [13]
	Cis-2-pentenal	*Tetraselmis* sp.; *Rhodomonas* sp.; *Nannochloropsis oculata; Botryococcus braunii; Chlorella*	Photoautotrofic	Durme et al., 2013 [8]
	2-Methylpropanal	*Phormidium autumnale*	Heterotrophic	Santos et al, 2015 [13]

Table 1. (Continued)

Organic classes	Compounds	Microalgae	Cultivation	Reference
	Hexanal	*Tetraselmis* sp.; *Rhodomonas* sp.; *Botryococcus braunii; Nannochloropsis oculata; Chlorella;*	Photoautotrofic	Durme et al., 2013 [8]
		Phormidium autumnale	Heterotrophic	Santos et al., 2015 [13]
	2-Methylbutanal	*Phormidium autumnale*	Heterotrophic	Santos et al., 2015 [13]
	Trans-2-hexenal	*Nannochloropsis oculata; Botryococcus braunii; Chlorella vulgaris*	Photoautotrofic	Durme et al., 2013 [8]
	Cis-4-heptenal	*Rhodomonas* sp.; *Botryococcus braunii*	Photoautotrofic	Durme et al., 2013 [8]
	Heptanal	*Rhodomonas* sp.; *Botryococcus braunii*	Photoautotrofic	Durme et al., 2013 [8]
	Tr,tr-2,4-heptadienal	*Rhodomonas* sp.; *Botryococcus braunii*	Photoautotrofic	Durme et al., 2013 [8]
	2-Octenal	*Nannochloropsis oculata; Botryococcus braunii*	Photoautotrofic	Durme et al., 2013 [8]
	Nonanal	*Rhodomonas* sp.; *Nannochloropsis oculata; Botryococcus braunii; Chlorella vulgaris*	Photoautotrofic	Durme et al., 2013 [8]
	2-Methylbutanal	*Phormidium autumnale*	Heterotrophic	Santos et al., 2015 [13]
	3-Methylbutanal	*Tetraselmis* sp.; *Rhodomonas* sp.; *Nannochloropsis oculata; Chlorella vulgaris*	Photoautotrofic	Durme et al., 2013 [8]
	Furfural	*Rhodomonas;*	Photoautotrofic	Durme et al., 2013 [8]
	Benzaldehyde	*Tetraselmis* sp.; *Rhodomonas* sp.; *Nannochloropsis oculata; Botryococcus braunii; Chlorella vulgaris*	Photoautotrofic	Durme et al., 2013 [8]
	Phenylacetaldehyde	*Nannochloropsis oculata; Chlorella vulgaris*	Photoautotrofic	Durme et al., 2013[8]

Organic classes	Compounds	Microalgae	Cultivation	Reference
	4-Ethylbenzaldehyde	*Botryococcus braunii*	Photoautotrofic	Durme et al., 2013 [8]
	Acetaldehyde	*Phormidium autumnale*	Heterotrophic	Santos et al, 2015 [13]
	Butanal	*Phormidium autumnale*	Heterotrophic	Santos et al., 2015 [13]
	2-Methylbutanal	*Phormidium autumnale*	Heterotrophic	Santos et al., 2015 [13]
	3-Methylbutanal	*Phormidium autumnale*	Heterotrophic	Santos et al., 2015 [13]
	Isobutyraldehyde	*Phormidium autumnale*	Heterotrophic	Santos et al., 2015 [13]
Hydrocarbons	Octane	*Phormidium autumnale*	Heterotrophic	Santos et al. 2015 [13]
	1-Heptene	*Phormidium autumnale*	Heterotrophic	Santos et al., 2015 [13]
	Hexadecane	*Spirulina platensis; Nostoc* sp.; *Anabaena* sp.	Photoautotrofic	Milovanovic et al., 2015 [38]
	Tetradecane	*Spirulina platensis; Nostoc* sp.; *Anabaena* sp.	Photoautotrofic	Milovanovic et al., 2015 [38]
	8-Methylheptadecane	*Spirulina platensis; Nostoc* sp.; *Anabaena* sp.	Photoautotrofic	Milovanovic et al., 2015 [38]
	3-Octadecene	*Nostoc* sp	Photoautotrofic	Milovanovic et al., 2015 [38]

Microalgae contain high concentrations of carotenoids, that by having a structure instable double bonds conjugated is easily degraded, for example, they can use the β-carotene which can degrade forming β-ionone and β-cyclocitral. In later stages of degradation of the β-ionone they can oxidize, forming a degradation product called 5,6-epoxy-β-ionone [15, 16].

As for the fatty acid synthesis, it occurs from the Acetyl-CoA molecule by Acetyl-CoA reductase enzyme. Using saturated fatty acids C_{18} as linoleic and linolenic acids, via the lipoxygenase 9-hydroperoxy form and intermediate 13-hydroperoxide. The branch hydroperoxide lyase converts both hydroperoxides C_6 and C_9 aldehydes such as 1-hexanal, hexanol and nonanal, which are reduced to alcohols by dehydrogenases [13]. In this synthesis may also be produced unbranched hydrocarbons by two families of enzymes: an acyl-acyl carrier protein reductase (AAR) and an aldehyde decarbonylase (AAD) that operate in the conversion of fatty acids [17].

A range of compounds, including classes, such as aldehydes, alcohols and ketones can be formed from the lipid degradation [7]. Microalgae are relatively rich in polyunsaturated fatty acids (PUFAs). Marine microalgae contain mostly very long chain PUFAs such as eicosapentaenoic acid and docosahexaenoic acid, for example, *Chlorella* contains principally shorter PUFA, such as α-linolenic acid. Species with low concentrations of PUFA contain a significantly smaller number of linear aldehydes compared with the species having high concentrations of PUFAs (e.g., *Chlorella*, *Botryococcus*, *Rhodomonas*) [18]. Considering that short chain linear aldehydes are often chemically derived lipid oxidation, branched aldehydes and aromatics are typically formed because of lipid oxidation and enzymatic protein [8].

Some chemical reactions may convert the volatile organic compounds into other compounds. For example, alcohols can be oxidized to aldehydes and then to carboxylic acids, and ketones may be reacted with the hydroxyl radicals in the air to form aldehydes [19, 20]. Aldehydes and ketones can be reduced to the alcohols by reductases aldehyde/ketone alcohols can be oxidized to aldehydes by alcohol dehydrogenase, and then further oxidized to the acid by aldehyde dehydrogenase [20]. Ketones can be formed in many ways; aliphatic ketones can be lipid oxidation products or ketone and methyl degradation (C_3-C_{17}) could be formed from the oxidative cleavage of carotenoids [13, 21].

Given the above, it is possible to note that some compounds can be produced metabolically (by enzymes present in microalgae) and also by primary degradation compounds such as lipids and proteins. The establishment of biochemical pathways can target specific biomolecules production of microalgae metabolism to compounds of commercial interest and also to better knowledge of their ecological function.

ECOLOGICAL IMPLICATIONS

The term "off-flavor" is used to describe the accumulation of odorous compounds within water or tissue produced from biological origins. This is one of the undesirable environmental implications, taste and odour outbreaks were associated with volatile organic compounds such as 2-methylisoborneol (2-MIB) and geosmin, produced by microalgae, are typical of the flavor compounds [14].

Geosmin is a bicyclic tertiary alcohol presenting earth odor even in very dilute aqueous solutions and it can be found naturally in beet and some plant roots. The metilisoborneol or 2-MIB also belong to the same chemical class of geosmin. Both are considered as semivolatile compounds terpenoids, being highly odorous in water or fish [22]. The 2-MIB and geosmin biosynthesis by microorganisms occurs by two common pathways: the mevalonic acid (MEV) and the deoxixilulose (DOXP/MEP) [23].

Geosmin and 2-MIB (Figure 2) are produced by aquatic microorganisms found in source waters such as lakes, reservoirs, and running waters. In addition, there are several other biological sources that are often overlooked, notably those which originate from terrestrial ecosystems, industrial waste treatment facilities, and drinking water treatment plants [24].

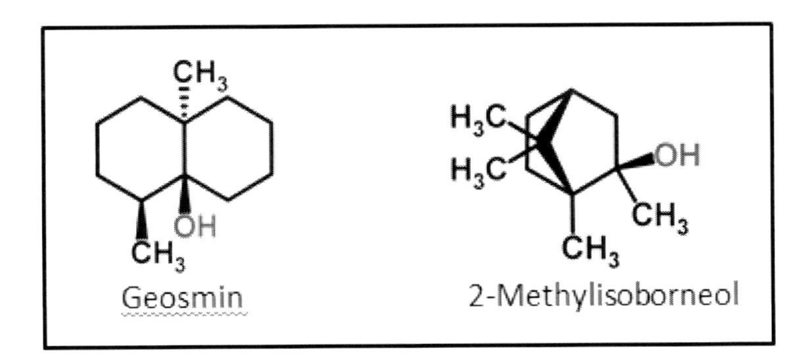

Figure 2. Chemical structure of geosmin and 2-methylisoborneol obtained from microalgae [50].

Microalgae are considered the main sources of geosmin and 2-MIB in aquatic environments where the photosynthetic growth is possible. These are present in freshwater lakes often form dense plankton populations or water blooms in eutrophic waters. In tropical regions, the growth of microalgae can be continuous throughout the year. Unsightly and highly visible surface blooms are usually considered to be primary sources of source water odor [24].

The formation of water blooms results from the redistribution, and often, rapid accumulation of buoyant planktonic populations. When such populations are subjected to optimal conditions, they respond by increasing their buoyancy and move upward nearer to the water surface, causing change in color of the water and often also in taste and odor.

The main reasons for the increased incidence of microalgae in water sources are: The increase in nitrogen nutrient loading and phosphate in water, which cause eutrophication of aquatic environments leading to an artificial enrichment of ecosystems. When this occurs in a relatively contained waterbody, there is an excessive proliferation of algae, due to decomposition, leading to an increased number of microorganisms, and thus, deterioration of water quality; In anaerobic medium inorganic forms of N and P predominate and facilitate uptake by microalgae, causing their blooms; The increase in organic matter load released springs directly or indirectly causes an increase in the amount of decomposing microorganisms and other sediment that eventually consuming the available oxygen in the water; Most microalgae blooms that appear in the springs consists of a few genres and usually produce toxins.

Apart from geosmin and 2-MIB, microalgae release other volatile organic compounds, which are also considered off-flavors, i.e., hydroxyketones formed by fermentation pathways, and carotenoids (e.g., β-cyclocitral) resulting from the degradation of carotenoids [25]. β-Cyclocitral is a well-known odor compound that affects drinking water supplies, and gives *Microcystis* blooms a characteristic hay tobacco odor, but its role in aquatic chemical defense against grazers has only recently been examined [26].

Table 2 describes some of off-flavors produced by some known species of microalgae. Microalgae, particularly filamentous, produce more than 25% of all known off-flavor compounds [27].

In the other hand, there is the possibility of synthesis of biogenic organic compounds by microalgae. In general, the term biogenic volatile organic compounds include organic atmospheric trace gases other than carbon dioxide and monoxide [28]. Consequently, large numbers of compounds saturated, unsaturated, and oxygenated are included within VOCs. And these are the isoprenoids (isoprene and monoterpenes), as well as alkanes, alkenes, carbonyls, alcohols, esters, ethers, and acids.

Table 2. Microalgae species known to produce off-flavor compounds

Source	Odorous metabolite(s)	Reference
Anabaena crassa	Geosmin	Watson (2003) [27]
Anabaena lemmermanii	Geosmin	Watson (2003) [27]
Aphanizomenon flos-aquae	Geosmin	Jüttner et al., (1986) [41]
Aphanizomenon gracile Lemmermann	Geosmin	Jüttner et al., (1986) [41]
Hyella sp.	MIB	Izaguirre and Taylor (1995)[42]
Leibleinia subtilis	Geosmin	Schrader and Blevins (1993) [43]
Lyngbya cryptovaginata	Geosmin	Jüttner and Watson (2007) [24]
Oscillatoria amphibia	Geosmin	Jüttner and Watson (2007) [24]
Oscillatoria limosa	MIB	Izaguirre and Taylor (1995) [42]
Odontamblyopus tenuis	MIB	Izaguirre et al., (1982) [44]
Phormidium amoeneum	Geosmin	Tsuchiya et al., (1981) [45]
Phormidium breve	Geosmin, MIB	Naes et al., (1988) [46]
Phormidium calcicola	Geosmin, MIB	Jüttner and Watson (2007) [24]
Phormidium. formosum	Geosmin	Persson (1988)[47]
Phormidium tenue	MIB	Persson (1988) [47]
Phormidium sp.	Geosmin, MIB	Zimmerman et al., (1995) [48]
Porphyrosiphon martensianus	MIB	Izaguirre and Taylor (1995) [42]
Rivularia sp.	Ketones, ionones	Höckelmann and Jüttner (2005) [49]
Tolypothrix distorta	Ketones, ionones	Höckelmann and Jüttner (2005) [49]

Isoprene and monoterpenes, in particular, as well as their reaction products are involved in tropospheric chemistry, fueling (directly or indirectly) the production of air pollutants and greenhouse gases, such as ozone, carbon monoxide, and methane, and increasing acidity as well as the production of aerossol [28, 29]. Usually these compounds are strong smelling, hardly water soluble, and found in plants as well as in animals, microorganisms as well as animals, microorganisms and microalgae [29]. These biogenic compounds serve as defense mechanisms of these microorganisms.

The group of monoterpenes comprises acyclic, and mono-, bi-, and tricyclic structures; they may exist as hydrocarbons with or without the inclusion of oxygen in compounds such as menthol, camphor, linalool, and geraniol. Oxygenated monoterpenes and their derivatives are often summarized as monoterpenoids [28]. Some examples of the dominant biogenic isoprenoids are given in Figure 3.

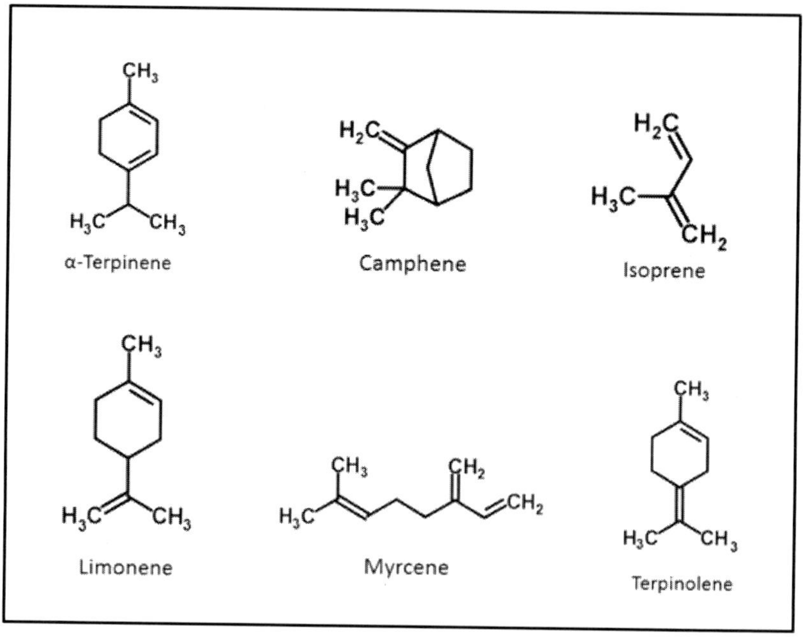

Figure 3. Dominant biogenic isoprenoids in microalgae [50].

In addition to metabolites that result in undesirable flavors and odors (odorous metabolites), there are those who are biochemically active (bioactive metabolites) in fresh and marine waters [28].

Finally, microalgae can also produce a wide range of volatile organic compounds (VOCs) and these compounds have diverse origins biosynthetic. Therefore, some of these odorous compounds apparently are the result of cell decay and decomposition. First, the algal cultures under investigation produced odors reminiscence of mercaptans and other organic sulfur compounds. Second, in the study of decaying cultures, one may logically argue that organic sulfur compounds may be present as a result of anaerobic decomposition of cellular material [30]. Decomposition products of any group are highly dependent on environmental conditions, especially temperature and oxygen available [31].

INDUSTRIAL APPLICATIONS

The most important product of microalgae biotechnology in relation to amount of production and economic value is its biomass. However, it has been noted an emerging trend towards knowledge production of low molecular weight compounds from renewable sources [32, 33].

Typical applications of microalgae correspond to a variety of metabolites (enzymes, lipids, biomass, pigments) with potential application in products such as cosmetics, food ingredients, and bioenergy. They can also be used as environmental indicators and for the treatment of wastewater [34, 35]. Beside many beneficial properties, microalgae also produce numerous volatile organic compounds, which could be used as an important alternative source of bulk and fine chemicals.

Volatile organic compounds generated by microorganisms have long been regarded as a breakthrough in laboratory research. Compounds with commercial appeal include propanol, butanol, 3-methyl-butanol, hexanol, hexanal, β-cyclocitral, β-ionone, and 5,6-epoxy-β-damascenone [36, 37].

Berger [36] reported that flavours from microorganisms can compete with traditional sources. The screening for overproducers, elucidation of metabolic pathways and precursors and application of conventional bioengineering has resulted in a set of more than 100 commercial aroma chemicals derived via biotechnology. Figure 4 shows the chemical structures of the commercialized compounds obtained by microorganisms and which are synthesized by microalgae showing a potential commercial application.

Generally, for each microalga species, aldehydes proved to be the most prevalent and, due to their low odor threshold values, might be important headspace volatiles compounds contributing to desirable aromas as well as rancid odors and flavors. Saturated aldehydes have a green-like, hay-like, paper-like odor, whereas unsaturated aldehydes have a fatty, oily, frying odor. Whereas the shorter chain linear aldehydes are often derived from chemical lipid oxidation, branched and aromatic aldehydes are typically formed due to enzymatic lipid and protein oxidation.

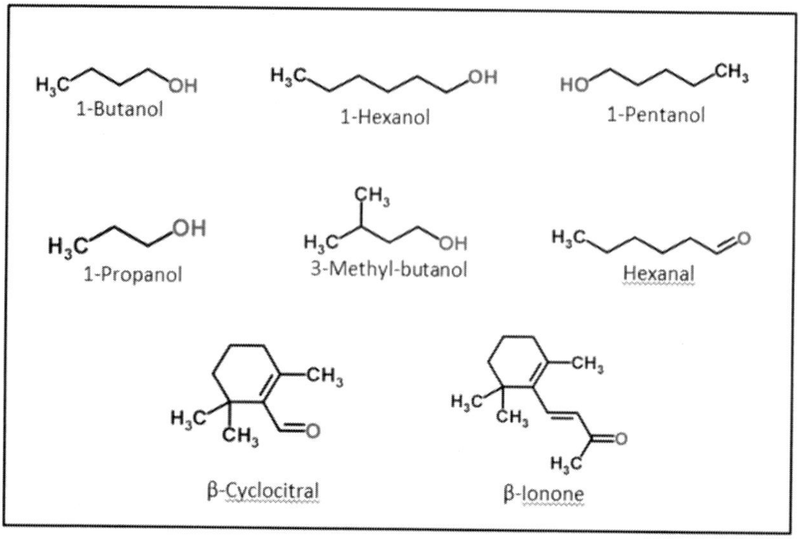

Figure 4. Volatile organic compounds with commercial application obtained from microalgae [50].

Many microalgae show the presence of ketones and alcohols as volatile compounds [8]. The volatile compounds determination shows that medium length alkanes and alkenes represent the main volatile components of the investigated strains of microalgae [38].

The full use of the volatile fraction of microalgal biomass may represent an improvement in the supply of a large volume of inputs to many different types of industry [13]. It can also occur using energy biomolecules of interest, such as hydrocarbons and short chain alcohols. There is increasing interest in the production of biofuels from renewable sources offering sustainable solutions to the energy sector as a promising alternative to traditional petrochemical industry [39].

The production of hydrocarbons is of particular interest due to their potential for use as advanced biofuels. Long-chain compounds can replace diesel, as the short-chain might do to instead of gasoline [33].

Aliphatic alcohols with higher carbon chain length or equal to five are attractive targets for biofuels have a high energy density and low water solubility (e.g., 1-pentanol 23 g/L; 1-hexanol 6.2 g/L; 1-heptanol 1.2 g/L). The enzyme responsible for the production of such compounds is the Acetyl-CoA-reductase that may be present in the reactions of the tricarboxylic acid cycle, mevalonate, and leucine biosynthesis. Other alcohol having substantial energy interest is the 1-butanol to have a comparable gasoline energy (29.2 MJ/L and 32.5 MJ/L, respectively), this can be a substitute fuel or added in the place of ethanol It has a lower energy (21.2 MJ/L) [40].

In summary, microalgae can produce a variety of industrially relevant volatile compounds, and the knowledge about the biosynthesis of these structures from microalgae might prove useful to help elucidated ways to the application of these biobased feedstocks for both food and non-food industries. In view of this commercial significance, efforts should be made to consolidate the technological routes of the production of these compounds.

References

[1] Buono, S.; Langellotti, A.L.; Martello, A.; Rinna, F.; Fogliano, V. (2014). Functional ingredients from microalgae. *Food & Function* 5, 1669-1685.

[2] Jacob-Lopes, E.; Scoparo, C.H.G.; Queiroz. M.I.; Franco, T.T. (2010). Biotransformations of carbon dioxide in photobiorreactors. *Energy Conversion and Management* 51, 894-900.

[3] Jacob-Lopes, E.; Franco, T.T. (2013). From oil refinery to microalgal biorefinery. *Journal of CO_2 Utilization* 2, 1-7.

[4] Nuccio, J.; Seaton, P.J.; Kieber, K.J. (1995). Biological production of formaldehyde in the marine environmental. *Limnology & Oceanography* 40, 521-527.

[5] Fay, P. (1983). *The blue-greens (Cyanophyta – Cyanobacteria)* (5th Ed.), Great Britain.

[6] Francisco, E.C.; Franco, T.T.; Wagner, R.; Jacob-Lopes, E. (2014). Assessment of different carbohydrates as exogenous carbon source in cultivation of cyanobacteria. *Bioprocess and Biosystems Engineering* 1, 2-11.

[7] Rzama, A.; Benharref, A.; Arrreguy, B.; Dufourc, E.J. (1995). Volatile compounds of green microalgae grown on reused wastewater. *Phytochemistry* 38, 1375-1379.

[8] Durme, J.V.; Goiris, K.; Winne, A.; De Cooman, L.; Muylaert, K. (2013) Evaluation of the volatile composition and sensory properties of five species of microalgae. *Journal of Agricultural and Food Chemistry* 61, 10881-10890.

[9] Kumar, K.; Dasqupta, C.N.; Nayak, B.; Lindblad, P.; Das, D. (2011). Development of suitable photobioreactors for CO2 sequestration addressing global warming using green algae and cyanobacteria. *Bioresource Technology* 102, 4945-4953.

[10] Iverson, T.M. (2006). Evolution and unique bioenergetic mechanisms in oxygenic photosynthesis. *Current Opinion in Chemical Biology* 10, 91-100.

[11] Papaleo, M.C; Romoli, R.; Bartolucci, G.; Maida, I. (2013). Bioactive volatile organic compounds from Antartic (sponges) bacteria. *New Biotechnology* 30, 824-838.

[12] Dudareva, N.; Klempien, A.; Muhlemann, J. K.; Kaplan, I. (2013) Biosynthesis, function and metabolic engineering of plant volatile organic compounds. *New Phytologist* 198, 16-32.

[13] Santos, A.B.; Fernandes, A.F.; Wagner, R.; Jacob-Lopes, E.; Zepka, L.Q. (2015). Biogeneration of volatile organic compounds produced by Phormidium autumnale in heterotrophic bioreactor. *Journal of Applied Phycology* 1, 1-10.

[14] Fujise, D.; Tsuji, K.; Fukushima, N.; Kawai, K.; Harada, K. (2010). Analytical aspects of cyanobacterial volatile organic compounds for investigation of their production behavior. *Journal of Chromatography A* 1217, 6122-6125.

[15] Rodrigues, D.B.; Flores, E.M.M.; Barin, J.S.; Mercadante, A.Z.; Jacob-Lopes, E.; Zepka, L.Q. (2014). Production of carotenoids from microalgae cultivated using agroindustrial wastes. *Food Research International*, 65, 144-148.

[16] Mendes-Pinto, M.M. (2009). Carotenoid breakdown products – the norisoprenoids – in wine aroma. *Archives of Biochemistry and Biophysics* 483, 236–245.

[17] Ducat, D.C.; Way, J.C.; Silver, P.A. (2011). Engineering cyanobacteria to generate high-value products. *Trends in Biotechnology* 29, 95–103.

[18] Zhang, Z.; Li, T.; Wang, D.; Zhang, L.; Chen, G. (2009). Study on the volatile profile characteristics of oyster Crassostrea gigas during storage by a combination sampling method coupled with GC/MS. *Food Chemistry* 115, 1150–1157.

[19] Atkinson, R.; Tuazon, E.C.; Aschmann, S.M. (2000). Atmospheric chemistry of 2-pentanone and 2-heptanone. *Environmental Science & Technology* 34, 623-631.

[20] Korpi, A.; Järnberg, J.; Pasanen, A.L. (2009). Microbial volatile organic compounds. *Critical Reviews in Toxicology* 39, 139-193.

[21] Sun, S.M.; Chung, G.H.; Shin, T.S. (2012). Volatile compounds of the green alga, Capsosiphon fulvescens. *Journal of Applied Phycology* 24, 1003-1013.

[22] Guttman, L.; Rijn, J.V. (2008). Identification of conditions underlying production of geosmin and 2-methylisoborneol in a recirculating system. *Aquaculture* 279, 85–91.

[23] Wanke, M.; Skorupinska-Tudek, K.; Swiezewska, E. (2001). Isoprenoid biosynthesis via 1-deoxy-D-xylulose 5-phosphate/2-C-methyl- D-erytritol 4-phosphate (DOXP/MEP) pathway. *Acta Biochimica Polonica Warsaw* 48, 663-672.

[24] Jüttner, F.; Watson, S.B. (2007). Biochemical and Ecological Control of Geosmin and 2-Methylisoborneol in Source Waters. Minireview. *Applied and Environmental Microbiology* 73, 4395–4406.

[25] Smith, J.L.; Boyer, G.L.; Zimba, P.V. (2008). A review of cyanobacterial odorous and bioactive metabolites: impacts and management alternatives in aquaculture. *Aquaculture* 280, 5–20.

[26] Watson, S.B.; Jüttner, F.; Köster, O. (2007). Daphnia behavioural responses to taste and odour compounds: ecological significance and application as an inline treatment plant monitoring tool. *Water Science & Technology* 55, 23–31.

[27] Watson, S.B. (2003). Cyanobacterial and eukaryotic algal odour compounds: signals or by-products? A review of their biological activity. *Phycologia* 42, 332–350.

[28] Kesselmeier, J.; Staudt, M. (1999). Biogenic Volatile Organic Compounds (VOC): An Overview on Emission, Physiology and Ecology. *Journal of Atmospheric Chemistry* 33, 23–88.

[29] Graedel, T.E. (1979). Terpenoids in the atmosphere. *Reviews of Geophysics and Space Physics* 17, 937–947.

[30] Jenkins, D.; Medsker, L.L.; Thomas, J.F. (1967). Odorous Compounds in Natural Waters. Some Sulfur Compounds Associated with Blue-Green Algae. *Environmental Science and Technology* 9, 731-735.

[31] Cheremisinoff, P.N. (1993). *Air Pollution Control and Design for industry*. Marcel Dekker Inc.

[32] Schirmer, A.; Rude, M.A.; Li, X.; Popova, E.; Cardayre, S.B. (2010). Microbial Biosynthesis of Alkanes. *Science* 329, 559-562.

[33] Choi, J.Y.; Lee, S.Y. (2013). Microbial production as short-chain alkane. *Nature*, 2013, 1-6.

[34] Jacob-Lopes, E.; Lacerda, L.M.C.F.; Franco, T.T. (2008). Biomass production and carbono dioxide fixation by Aphanothece microscopica

Nagëli in a bubble column photobioreactor. *Biochemical Engineering Journal* 40, 27-34.

[35] Abdel-Raouf, N.; Al-Homaidan, A.A.; Ibraheem, I.B.M. (2012). Microalgae and wastewater treatment. *Saudi Journal of Biological Science* 19, 257-275.

[36] Berger, R.G. (2009). Biotechnology of flavours - the next generation. *Biotechnology Letters* 31, 1651-1659.

[37] Smith, K.M.; Cho, K.M.; Liao, J.C. (2010). Engineering Corynebacterium glutamicum for isobutanol production. *Applied Microbiology and Biotechnology* 87, 1045-1055.

[38] Milovanovic, I.; Mison, A.; Simeunovic, J.; Kovac, D.; Jambrec, D.; Mandic, A. (2015). Determination of volatile organic compounds in selected strains of cyanobacteria. *Journal of Chemistry*, 2015, 1-6.

[39] Si, T.; Luo, Y.; Xiao, H.; Zhao, H. (2014). Utilizing an endogenous pathway for 1-butanol production in Saccharomyces cerevisiae. *Metabolic engineering* 22, 60-68.

[40] Zhang, K.; Sawaya, M. R.; Eisenberg, D. S.; Liao, J. C. (2008). Expanding metabolism for biosynthesis of non-natural alcohols, *Proceedings of the National Academy of Sciences*, 105, 20653-20658.

[41] Jüttner, F.; Hoflacher, B.; Wurster, K. (1986). Seasonal analysis of volatile organic biogenic substances (VOBS) in freshwater phytoplankton populations dominated by Dinobryon, Microcystis and Aphanizomenon. *Journal of Phycology* 22, 169–175.

[42] Izaguirre, G.; Taylor, W.D. (1995). Geosmin and 2-methylisoborneol production in a major aqueduct system. *Water Science & Technology* 31, 41–48.

[43] Schrader, K.K.; Blevins, W.T. (1993). Geosmin-producing species of Streptomyces and Lyngbya from aquaculture ponds. *Canadian Journal of Microbiology* 39, 834–840.

[44] Izaguirre, G.; Hwang, C.J.; Krasner, S.W.; McGuire, M.J. (1982). Geosmin and 2- methylisoborneol from cyanobacteria in three water supply systems. *Applied and Environmental Microbiology* 43, 708–714.

[45] Tsuchiya, Y.; Matsumoto, A.; Okamoto, T. (1981). Identification of volatile metabolites produced by blue-green algae Oscillatoria splendida, Oscillatoria amoena, Oscillatoria geminata and Aphanizomenon sp. *Yakugaku Zasshi Journal of the Pharmaceutical Society of Japan* 101, 852–856.

[46] Naes, H.; Utkilen, H.C.; Post, A.F. (1988). Factors influencing geosmin production by the cyanobacterium Oscillatoria brevis. *Water Science & Technoogy* 20, 125–131.

[47] Persson, P.E. (1988) Odorous algal cultures in culture collections. *Water Science & Technology* 20, 211–213.

[48] Zimmerman, W.J.; Soliman, C.M.; Rosen, B.H. (1995) Growth and 2-methylisoborneol production by the cyanobacterium Phormidium LM689. *Water Science & Technology* 31, 181–186.

[49] Höckelmann, C.; Jüttner, F. (2005). Off-flavours in water: hydroxyketones and β-ionine derivatives as new odour compounds of freshwater cyanobacteria. *Flavour and Fragrance Journal* 20, 387–394, 2005.

[50] Royal Society of Chemistry (2015). *ChemSpider: search and share chemistry* (http://www.chemspider.com/). Access on 20 February 2015.

In: Volatile Organic Compounds
Editor: Julian Patrick Moore

ISBN: 978-1-63485-370-5

Chapter 2

VOLATILE ORGANIC COMPOUNDS FROM INDUSTRIAL COMPLEXES AND WASTEWATER TREATMENT PLANTS: OCCURRENCES, BEHAVIORS AND HEALTH RISKS

Wei-Hsiang Chen *

Institute of Environmental Engineering, National Sun Yat-sen University, Kaohsiung, Taiwan

ABSTRACT

The emission of volatile organic compounds (VOCs) from industrial stationary sources is becoming an environmental issue of increasing concern. Industrial complexes and wastewater treatment plants (WWTPs) represent two important VOC sources from industrial activites. This chapter attempted to investigate the VOC pollutions from industrial complexes and WWTPs by starting from summarizing selected articles and discussing the effects of altitude and tropical savanna climate on the VOC distributions in Kaohsiung City of southern Taiwan. Furthermore, additional articles that investigated the VOC emission rates, mass distirbutions among different phases (e.g., air, water, and sludge), and the associated adverse health risks to local workers and populations living

* Corresponding Author address. Email: whchen@mail.nsysu.edu.tw.

near a WWTP were selected for discussion. In the first discussion, the the altitude effect on VOC distributions was introduced by analyzing the VOC concentrations at ground level and three different altitudes and assessing the results by principal component analysis (PCA) and cluster analysis. Next, the influences of diurnal temperature and seasonal humidity variations by tropical savanna climate on the distributions of VOCs were investigated. For the VOC emissions in WWTPs, biological treamtent is the process more likely to be relevant to the VOC productions and modified to mitigate the VOC emissions. By conducting lab-scale batch simulation experiments, this capter futher discussed the fates of aromatic and chlorinated hydrocarbons in wastewater treatment processes, with respect to the difference amongst the species, together with the effects of aeration and the presence of activated sludge. With the growing concern regarding the VOC emissions from WWTPs, the relationship between the VOC emission rates and the associated public health risks was also compared and examined by using a municipal WWTP in northern China as an example. Last, the VOC distributions through the full-scale WWTP were fully investigated with respect to the effects of seasonal temperature and treatment technology variations, followed by the adverse health risk assessment that simulated the atmospheric behaviors of the VOCs emitted from the WWTP and calculated the associated cancer and non-cancer risks. The discussion here followed that the VOCs emitted from a WWTP could be one important concern for the chronic adverse health risks for both the workers in the WWTP and those people living some distances away from the plant. It emphasized the complex nature of VOC emissions from WWTPs and quantitatively indicated that the associated health impacts to the public near the WWTPs could be severely underestimated, whereas their treatment efficiencies by wastewater treatment technologies were overestimated.

Keywords: volatile organic compound (VOC), industrial complex, altitude effect, tropical Savanna climate, wastewater treatment plant (WWTP), health risk

INTRODUCTION

Volatile organic compounds (VOCs) are easily emitted to the atmosphere due to their high vapor pressures and low boiling points. These compounds are of concern for several reasons. First, many VOCs are important precursors or reactants that form secondary air pollutants such as ozone and photochemical smog [1]. Many VOCs are also hazardous air pollutants (HAPs) that pose environmental and public health risks [2]. More importantly, a number of VOCs cause chronic hazards of skin, central nervous system, liver, and kidney, while many of them are mutagenic, tetratogenic, or carcinogenic [3]. Besides natural sources such as vegetation and forest fires, VOCs are emitted from anthropogenic sources such as industrial and residential activities, vehicular emissions, and landfill. Due to their ubiquitous occurrences and adverse effects to human health, investigating the VOC emissions from various anthropogenic sources to minimize their adverse impacts on the environment and human health is becoming important issues in these years. This chapter attempted to review and summarize the findings of several articles that investigated the VOC emissions from two imporant anthropogenic sources including industrial complexes and wasetwater treatment plants (WWTPs) with respect to differetn issues. These issues include the influneces of altitude, tropical savanna climate, and characeteristics of molecular strucuture and emission source on the VOC emissions and the associated adverse health risk assessments.

Temporal and spatial distributions of VOCs emitted from different sources have been the focus of many early studies. However, the information regarding the vertical distributions of VOCs with heights is rather limited. The VOCs at different heights in the atmosphere are affected by turbulent dispersion, transport, and mixing in the atmospheric environment and thus less effectively predicted by discrete monitoring events, as these samplings and analyses are typically conducted at or near the ground level. The type of emission source is another important factor that impacts the vertical distribution of VOCs. For example, the VOCs emitted from elevated flare stacks in industrial complexes or incinerators can reach higher altitudes and are more likely to be transported over long distances. In the 1st paper reviewed in the chapter [4], by focusing on the VOCs emitted from industrial complexes in Kaohsiung City of southern Taiwan, the VOC concentration distributions at different altitudes in the atmosphere were investigated, providing the valuable information regarding the altitude effect on the VOC distribution and influential factors with statistical approaches including principal component analysis (PCA) and cluster analysis.

Kaohsiung City is one of the areas of interest in the reviewed papers selected and discussed in this chapter. Although it is the largest industrial city in southern Taiwan with approximately three million populations, the city is known for its air pollution near the metropolitan areas [4, 5]. A dense population of different industries neighbors the north of the city and are known to deteriorate the ambient air quality in and near the areas with their intensive

VOC emissions. Another interest for the VOC pollution in this city is that the area is located within a degree to the Northern tropic and is affected by a tropical wet and dry climate, known as the tropical savanna climate. This climate is characterized by its high temperature in every month through the year and a pronounced dry season. As climate facotrs are expected to importantly affect the VOC distribution and transport in the atmosphere, the exact effect of strong temporal variations in the regions of tropical savanna climate on the temporal and spatial distributions of VOC pollution is limited known and imperative for characterizing the VOC behaviors for effective management. This chapter reviewed aother paper that monitored the VOC concentrations in the atmosphere in Kaohsiung City during the daytime and nighttime in the dry and wet seasons [5]. During the sampling periods in that study, the temperatures elevated during the daytime through the year with the occurrences of night-time lower temperatures, and the humidity substantially varied between the dry and wet seasons. Given these environmental conditions, the effects of diurnal temperature and seasonal humidity variations due to the tropical savanna climate on the distributions of anthropogenic VOCs were discussed and summarized in this chapter. The information helps understand the possible changes of the inherent characteristics behind different VOC distributions at varying temperatures and humidity levels.

Besides the industrial sources, the concern for the VOC emissions in WWTPs is also growing [6-9]. VOCs originally present in source wastewater are emitted into the environment via volatilization through different wastewater treatment technologies. Mandated by the Clean Air Act, the public WWTPs in the U.S. are required to inventory and diminish the emissions of VOCs and HAPs [2, 10]. Currently, the VOC emissions in WWTPs are under close scrutiny and investigation by public authorities and government agencies in many countries of the world. In typical WWTPs, biological processes defines one of the the best approaches to treat VOCs in wastewaters. However, concerns regarding the effectiveness and efficiency of VOC treatment in WWTPs arise as a consequence of the volatilization of VOCs in treatment processes [6, 8]. Both biological process such as activated sludge and volatilization are capable of reducing the VOC concentrations in wastewaters. Nevertheless, the impacts to the environment and public health may occur due to the potential of overestimating the VOC remvoal efficiency when appreciable fractions of VOCs are actually removed via volatilization into the environment [6, 8].

The emission rates of different VOCs in WWTPs are expected to be different, as the emission process is controlled by diffusion from water to air phases and is affected by many water quality and operational parameters including the concentration gradient between air and water phases, temperature, and type of treatment technology [9]. Additionally, the toxicity varies amongst different VOC species. The public health risks posed by the VOCs emitted from WWTPs may vary significantly by VOC species, treatment processes, source waters, or emission seasons [7, 9]. Previous studies have emphasized the need for investigating and evaluating the adverse health risks to public, including both carcinogenic and non-carcinogenic effects, due to the exposures to VOCs emitted from WWTPs [9, 11]. While the VOC emission from WWTPs have been the concern of many preceding studies, the findings of several articles that investigated the VOC emission rates, mass distirbutions among different phases (e.g., air, water, and sludge), and the associated adverse health risks to local workers and populations living in and near a WWTP were selected for discussion and summairized in this chaper [6-9].

METHODS

Monitoring and Sampling Areas

In the papers selected for discussion in this chapter, which invesitgated the effects of altitude [4] and tropical savanna climate [5] on VOC distributions and transports, several industrial complexes in Kaohsiung City of southern Taiwan was the monitoring and sampling areas to analyze the VOC concentrations in the atmosphere [4]. More detailed information describing the procedures of sampings and analyses is available in the assocated references. In the 2nd half of this chapter, the VOC emissions and the associated health risks in and near a WWTP is the research topic of interest. In the papers selected and summarized in thie chapter [6-9], a municipal WWTP located within 10 km of the downtown of Harbin City in northern China was selected. The total daily design treatment capacity of the WWTP is 325 thousand cubic meters per day. Ninety-five percents of the source wastewater received in the WWTP is domestic sewage, while the remaining 5% is from local industrial sources. With the complexity of source water quality and an apparent ambient temperature change among seasons at the WWTP located in Harbin City, the emission rates ad adverse health risks of different VOCs to the workers in the WWTP and public near the plant were investigated in the selected studies and discussed in this chapter. More detailed information that characterizes the

WWTP and describes the sampings and analyses is available in the assocated references [6-9].

Field Sampling

In the reviewed papers that investigated the effects of altitude and tropical savanna climate on the VOC distributions and transports, air samples were collected at ground level and different altitudes in three seasons. For each sampling, four different sites were selected to consider the variation of wind direction during the sampling periods. At least one upwind location and one downwind location were included for sampling. The Taiwan Air Quality Monitoring Network (TAQMN) was used to collect the wind direction information of the studied areas during the monitoring periods (http://taqm.epa.gov.tw/taqm/en/default.aspx). Tethered balloon technology was used to collect air samples at high altitudes [4, 5]. Simply speaking, the balloons with a pair of 10 L tedlar sampling bags hung on the tethered line right beneath the balloon were raised to selected altitudes. When the balloon reached the predetermined altitudes, the air sampling pump started to collect air sample. The balloon was sent back to the ground and the air sample in the bag were drawn into the specially-designed stainless steel canisters. The canister was transported to the laboratory, followed by analysis to determine the VOC concentrations in the air sample.

Bench-Scale Bioreactor Experiment

It is difficult to study the fates of the VOCs in real WWTPs in which the complexity by the combination of water quality variation and practical operational adjustment may blur the findings from the field experiments. Bench-scale experiments were undertaken in the papers selected for review in this chapter to simulate the wastewater treatment processes in the WWTP of interest, including the primary sedimentation, aerobic biological treatment, and secondary sedimentation [6-9]. The chemical oxygen demand (COD) of the source wastewater ranged from 282 to 487 mg/L during the sampling periods. The wastewater and activated sludge in the bioreactor experiments were collected from the selected WWTP. The reactor has a capacity of 27 L (0.3 m (L) x 0.3 m (W) x 0.3 m(H)) and three aeration pipes connected at the bottom for aeration. The processes in the bioreactor experiments included the primary sedimentation without aeration for 2 hours, followed by aerobic sludge treatment process with aeration for 8 hours and secondary sedimentation without aeration in the last 2 hours. The source water in the experiments was the effluent from the primary treatment processes of the WWTP, while the activated sludge from the full-scale activated sludge process was added into the bioreactor with a pre-determined mixed liquor suspended solid concentration (MLSS). The water was aerated with multiple rates to control the dissolved oxygen concentration in the water. Air, water, and sludge samples were collected at regular intervals to measure the VOC concentrations.

VOC Sampling and Analysis

For air sampling, if the wastewater was continuously disturbed by aeration (e.g., acitvated sludge), air samples were collected at 1 meter above the water surface. In the processes without aeration, samples were collected at the same position but at 0.1 m above the water surface. The QT-2B VOC sampler with an activated carbon-filled GH-1 adsorption tube was used to collect air samples. The VOCs adsorbed onto the adsorption tube were extracted by rinsing with dichloromethane for 12 hours. The extracted solutions were concentrated and stored at 4°C, followed by the analysis in 72 hours. To analyze the VOC concentrations in the water phase, 5 ml of wastewater after filtration with 0.45 μm pore size membrane filters were collected and sealed in headspace vials. The VOC concentrations in the water phase were quatified by using the headspace gas chromatography coupled with mass spectrometry (GC/MS). Simply speaking, the water sample was placed in a closed vessel and heated at 80°C for 30 minutes. The vapor produced in the vessel was sampled and analyzed to determine the VOC concentrations in the wastewater. Sodium chloride was added in the vials to improve the sensitivity of the VOC analysis by increasing the ionic strength and volatilization of VOCs. All wastewater samples were stored at 4°C and analyzed in 72 hours after the sampling. To analyze the VOC concentrations in the sludge phase, 100 ml of wet sludge were collected. The sludge sample was slowly filtered with a 0.45 μm pore size membrane filter for 1 hour. Soxhlet extraction using dichloromethane as the solvent was used to extract the VOCs in the sludge retained on the filters. After the extraction for 24 hours, the extract was concentrated to 1 ml and prepared for the following GC-MS analysis within 72 hours.

The VOC concentrations in the papers summarized in this chapter were mainly analyzed by the following approach. Air samples were collected in specially-prepared canisters, which were emptied and cleaned before sampling. The canisters were transported to the laboratory after the sampling and the VOC concentrations were analyzed by using GC/MS (HP 6890N) and a thermal desorption unit (ENTECH 7100A, TDU). During analysis, air sample was directed from a canister through a solid multi-sorbent concentrator, and the moisture in the sample was removed by dry-purging with helium. The concentrator was heated at 150°C so the VOCs were desorbed and entrained in the carrier gas stream. The VOCs were trapped on a small volume multi-sorbent trap, thermally desorbed, and then injected into the GC/MS. For the VOC analyses in the water or extract samples, the samples were cleaned up and injected to determine the VOC concentrations in the solution phase. More information regarding the sampling and analytical methods is available in the associated references [5-7].

The Fugacity Model

The fugacity model is a common method used to describe the interphase-transfers of a compound in a multiple environmental compartments by predicting the potentials of transferring amongst the media with mathematical statements. The potential of a compound for escaping from any given phase is represented as fugacity and evaluated by contrasting its concentration to whatever concentration would be expected at equilibrium with the reference state. By knowing the fugacity of a compound in different environmental phases, the transfer of a compound in the system is characterized. A compound is expected to transfer to another phase if its fugacity in the original phase is greater. The extent of this transfer is proportional to the extent of the difference between its fugacity in two phases. The transfer of the compound continues until its fugacity equals in two interacting phases. The fugacity calculation considers the air phase as the reference state, as the fugacity in the water and sludge phases were calculated by relating the concentrations to their corresponding air-phase concentrations at equilibrium. The fugacity of a compound is typically represented in normalized pressures [12]:

$$\text{Fugacity of the compound in air} = C_{air}RT$$
$$\text{Fugcity of the compound in water} = C_{water}K_{aw}RT$$

$$\text{Fugacity of the compound in sludge} = \frac{C_{sludge}}{K_{sludge}} K_{aw} RT$$

where C_{water}, C_{air}, and C_{sludge} are the concentrations of a compound i in the air, water, and sludge phases, respectively [Mass/Length^3]; K_{aw} is the dimensionless Henry's law coefficient; K_{sludge} is the dimensionless partition coeffient between the sludge and water phases; and R and T are the gas constant [Pa-L/mol-K] and temperature [K], respectively. The partition coefficent of a compound i between the sludge and water phase was kinetically dominated by the fractions of lipid and protein in the sludge [12]:

$$K_{sludge} = f_{lipid} K_{lipid} + f_{protein} K_{protein}$$

where f_{lipid} and $f_{protein}$ represent the dimensioless fractions of lipid and protein in the sludge on a weight basis, respectively; and K_{lipid} and $K_{protein}$ denote the dimensionless partitioning constants between the lipid- and protein- to the water-phase concentrations, respectively. The K_{lipid} and $K_{protein}$ values of a compound i were derived by linear free energy relationships (LFERs) and its octanol-water partition constants (K_{ow}) [12]:

$$K_{ilipid} = 3.2 \times K_{ow}^{0.91}$$
$$\log K_{protein} = 0.7 \times \log K_{ow}$$

The information regarding the partitioning coefficients of VOCs used for calculation is collected from the database of Agency for Toxic Substances and Disease Registry (www.atsdr.cdc.gov).

Multivariate Statistical Analysis

PCA is a method that explores the variability in multivariate data. It has been widely used to investigate and study air pollution issues such as major air pollutants, influential atmospheric factors, or possible sources [4, 5, 13]. The air quality data typically contains a great number of observed and recorded data in discrete form, which is constituted by series of air pollutants and weather variables. An eigenvalue decomposition of the variance matrix of data is used to gain additional intuitions behind the environmental data and to find the directions in observations along which the data have the highest variability.

The results such as those correlated variables are converted into a set of linearly independent variables, namely principal components, which are represented as functions of the original variables and are considered as single observations in function space rather than as high dimensional vectors. After the PCA transformation, the first principal component has the largest variance to account for as much of the variability in observations as possible. The second component in turn has the largest variance uncorrelated with those of the previous components. In the papers selected and reviewed in this chpater, the profiles of VOC concentrations were analyzed by the PCA application, which was conducted with SPSS 17.0 to determine the principle components. In the process, the data were preliminarily standardized by calculating the unweighted average concentration of each VOC species with identical units to eliminate the effects of different sizes and units of measurement. By calculating the eigenvalues of the original variables, varimax orthogonal rotation was applied to determine the principal components of observations. As the Kaiser Criterion was used to determine the number of principal components, only the principal component with an eigenvalue exceeding 1 was considered. The loading values were used to represent the correlations between observations and components. A loading values larger than 0.7, between 0.5 and 0.7, and less than 0.5 indicate strong, medium, and weak correlations, respectively. More detailed information regarding the PCA method is available in the previously published studies and papers selected for review in this chapter [14].

The PCA results, which suggested the important VOC species under different circumstances, were further analyzed by the cluster analysis. The loading values of each VOC to the 1st, 2nd, and 3rd components were plotted so the structures of the data could be virtually examined and discussed. A model-based clustering algorithm, namely *mclust* using the statistical language R, was applied to effectively cluster the VOC species with similar characteristics [15]. Additional information regarding the *mclust* package used in these reviewed papers can be acquired by the following link: http://www.stat. washington.edu/mclust/.

Emission Rate Estimation

One of the papers selected in this chapter estimated the VOC emission rates in different treatment processes of a WWTP [9]. In the estimation, the treatment processes were divided into three categories: a close system, an open system, and an open system with disturbed water surface such as activated sludge. For a close system, the concept of mass balance was used to estimate the VOC emission rates [9]:

$$V\frac{dC_i}{dt} = k_0 q_0 C_0 - k_1 q_1 C_1 + S_{close}$$

where C_0 and C_i are the gas-phase VOC concentrations in the air flowing in and out of the system, respectively [Mass/Length^3]; q_0 and q_1 are the air flow rates in and out of the system, respectively [Length^3/Time]; t, V, and S_{close} are the time [Time], the gas-phase volume in the system [Length^3], and the emission rate of the VOC of interest [Mass/Time], respectively; and k_0 and k_1 represent the ratios of the air volume into and out of the system to the volume of the headspace in the system, respectively.

The bar screen in the WWTP was considered as a close system. No VOCs was present in the air into the process. The VOC emission rates in the bar screen were predicted by the equation below:

$$C_i = \left(\frac{q_0 C_0 + S_{close}}{k_1 q_1}\right)(1 - \exp(-\frac{k_1 q_1 t}{V}))$$

The primary and secondary sedimentation basins and anaerobic sludge treatment process in the WWTP were considered as an open system. The VOC emission rates in these processes were deyermined as follows [16]:

$$S_{open} = K_L A C_L$$

where S_{open} represents the VOC emission rate in an open system; A is the water surface area per unit volume of water [Length^2/Length^3]; C_L is the water-phase VOC concentration [Mass/Length^3]; and K_L is the overall mass transfer coefficient of the VOC between the air and water phases [Length/Time]:

$$\frac{1}{K_L} = \frac{1}{k_l} + \frac{1}{K_{aw} k_g}$$

where k_l is the dimensionless water-phase mass transfer coefficient of the VOC [Length/Time]. Given different wind speed conditions, the k_l value can be calculated by using the equations below:

$k_l = 2.78 \times 10^{-6}(\frac{D_{i,w}}{D_{ether,w}})^{2/3}$ when 0<U<3.25

$k_l = (2.605 \times 10^{-9}\left(\frac{F}{D}\right) + 1.277 \times 10^{-7})U(\frac{D_{i,w}}{D_{ether,w}})^{2/3}$ when 3.25<U and 14<F/D<51.2

$k_l = 2.611 \times 10^{-7}U^2\ (\frac{D_{i,w}}{D_{ether,w}})^{2/3}$ when 3.25 < U and F/D > 51.2

where $D_{i,\ w}$ and $D_{ether,\ w}$ denote the diffusion coefficients of the VOC of concern and ether, respectively [$Length^2/Time$]; F/D is the fetch-depth ratio; and U is the wind speed above the water surface [Mass/Time]. The dimensionless gas-phase mass transfer coefficient (k_g) of a VOC was estimated as follows [16]:

$$k_g = 4.82 \times 10^{-3}U_{10}^{0.78}Sc^{-0.87}de^{-0.11}$$

$$Sc = \frac{G}{\rho_{air}D_{i,a}}$$

where Sc is the Schmidt number of the VOC in the gas phase; de is the effective diameter of water surface [L]; G and ρ_{air} denote the viscosity and density of air, respectively; and $D_{i,a}$ is the diffusion coefficient of the VOC in air [$Length^2/Time$].

The activated sludge in the WWTP was considered as an open system with a disturbed water surface. The emission rates of VOCs in the process were enhanced by aeration. As a result, the emission rates were calculated by using the theory developed by Mattermuller et al. [17]. The theory can be summarized and described by the following equations [9]:

$$S = Q_G K_{eq}\left(1 - \exp\left({}^{K_L A}\!/_{K_{eq}Q_G}\right)\right)C_L$$

$$K_L = K_L^{oxy}\left({}^{D_{i,w}}\!/_{D_{oxy,w}}\right)$$

where Q_G is the air flow rate [$Length^3/Time$]; D_{oxy} is the molecular diffusion coefficient of oxygen in water [$Length^2/Time$]; H is the water depth [Length]; K_L^{oxy} is the overall mass transfer coefficient of oxygen [Length/Time]. Additional detailed information regarding the methods and certain parameters used but not mentioned is available in the associated reference [9].

Health Risk Assessment

Two papers were selected for discussion in this chapter, as these studies investigated the adverse health risks posed to the local workers and public near the WWTP due to the exposures to the VOCs emitted in the WWTP [7, 9]. Inhalation intake was the primary exposure pathway. The following equations were used to calculate the possible cancer risk of an individual likely to acquire cancer in the lifetime due to his or her exposure to VOCs from inhalation intake [3, 7, 9, 18].

$$\text{Cancer Risk} = \text{LADD} \times \text{SF}$$

$$\text{LADD} = \frac{\text{C} \times \text{InhR} \times \text{IAF} \times \text{ET} \times \text{LRF} \times \text{EF} \times \text{ED}}{\text{LT} \times \text{BW} \times 365\text{day/year}}$$

where LADD is the life time average dose [mg/kg-day]; SF is the slope factor [kg-day/mg]; C is the air-phase VOC concentration [mg/m^3]; InhR is the inhalation rate [m^3/hr]; IAF is the inhalation intake adjustment factor; ET is the average exposure time [hr/time]; LRF is the lung retention factor; EF is the average exposure frequency [time/year]; ED is the working exposure duration [years]; LT is the life time [year]; and BW is the body weight [kg].

The risk of an individual likely to acquire non-carcinogenic diseases by inhalation of VOCs was investigated by calculating the hazardous quotients (HQ).

$$\text{HQ} = \text{C/RfC}$$

where C is the air-phase VOC concentration by inhalation [mg/m^3] and RfC is the inhalation reference concentration [mg/m^3]. The RfC is a reference number that helps predict a concentration to people who are likely to be without risks of deleterious effects for their lifetime through a continuous inhalation exposure. A HQ larger than 1 suggests that a non-cancer risk occurs given the exposure to the levels of VOCs. The information regarding the slope factors and RfC numbers were collected from the Integrated Risk Information System (IRIS) by the USEPA or the California Environmental Protection Agency (CalEPA) [3, 19].

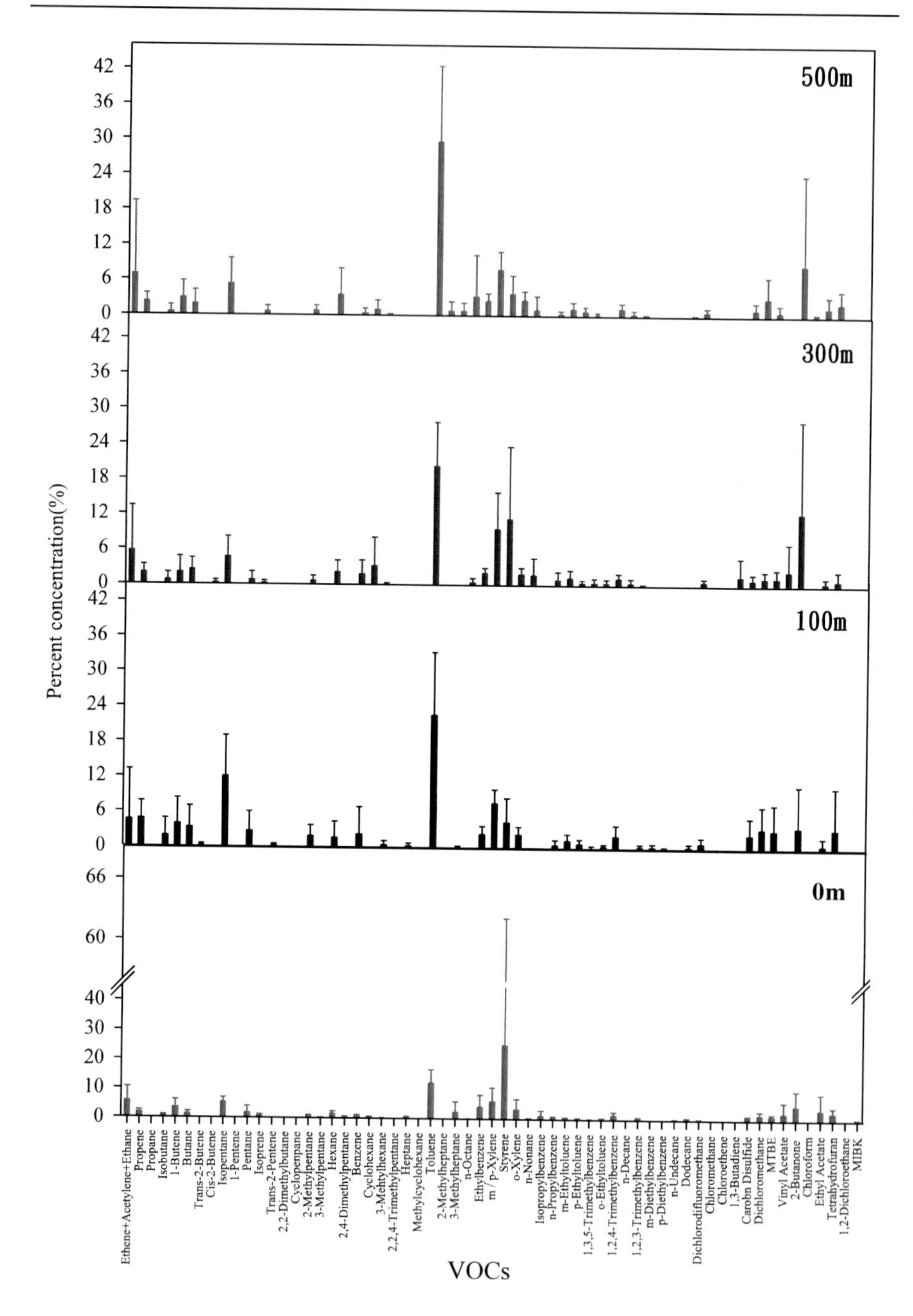

Figure 1. Concentrations of VOCs detected at monitoring sites near the selected industrial complexes in Spring of 2009 [4].

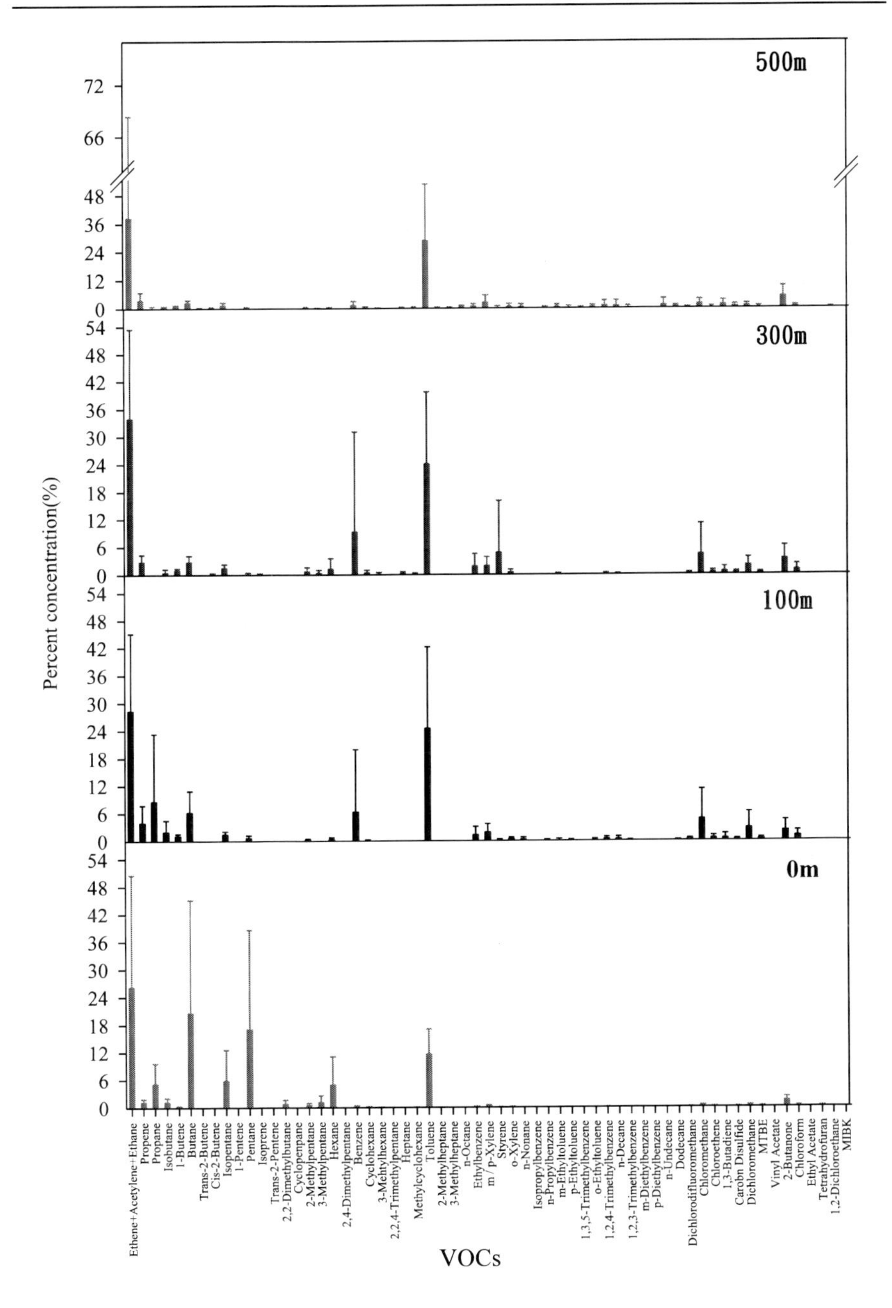

Figure 2. Concentrations of VOCs detected at monitoring sites near the selected industrial complexes in Winter of 2009 [4].

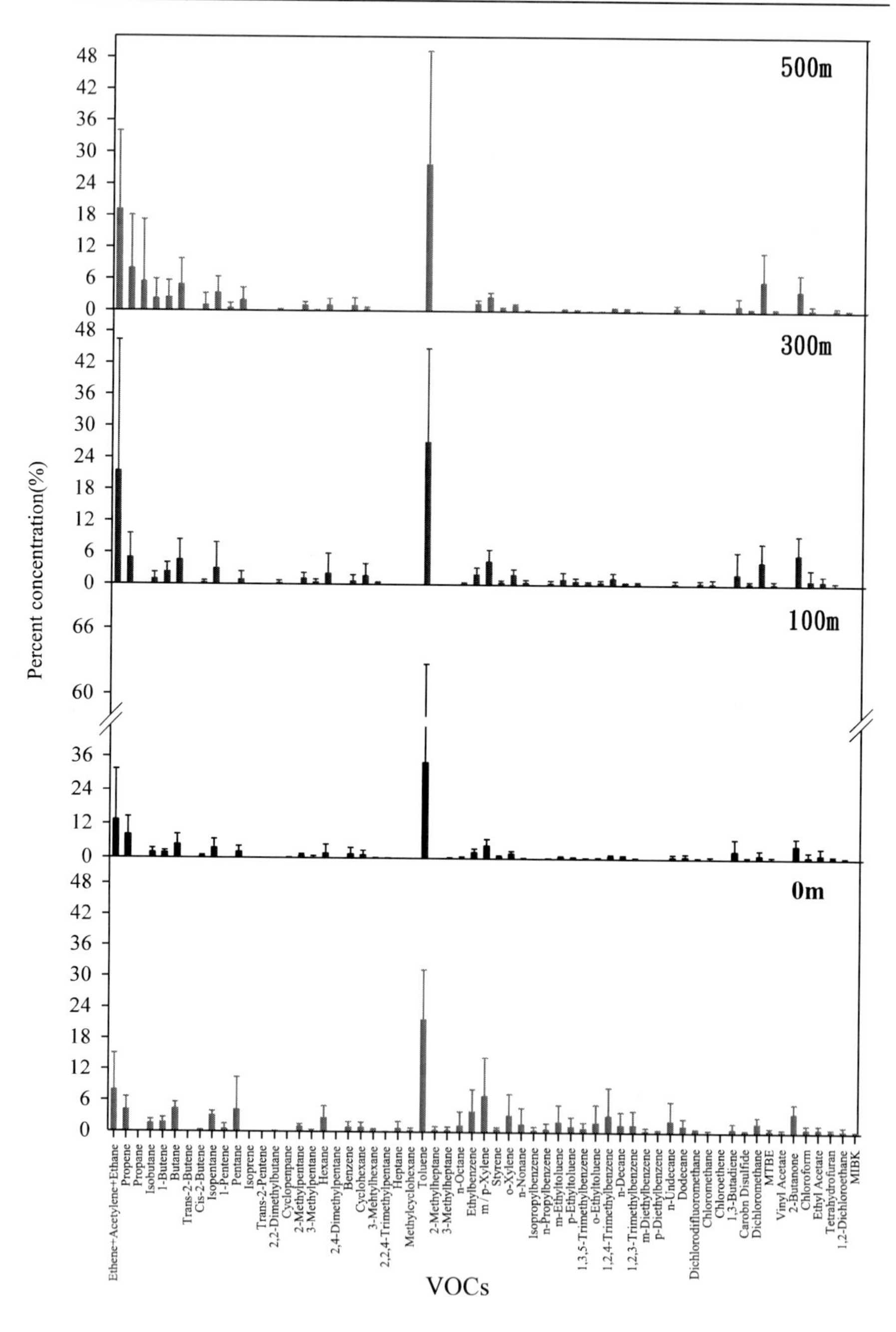

Figure 3. Concentrations of VOCs detected at monitoring sites near the selected industrial complexes in Spring of 2010 [4].

RESULTS AND DISCSSION

The Altitude Effect

The paper prepared by Yang et al. in 2013 studied the altitude effect on the distributions and transports of the VOCs emitted from the industrial complexes at different times and seasons. Shown in the paper [4], the concentration distributions of the VOCs at ground level and 100, 300, 500 m above the ground in three monitoring seasons are illustrated in Figures 1 through 3. Acetone and toluene were two major VOC species observed in the results, as the acetone concentration reached up to 400 ppb. Given its substantially high concentration, acetone and the other VOCs at negligible levels were not included in these figures. Many VOCs with moderate concentrations near ppb levels were observed. These VOCs included chloro- and dichloro-methane, propane and propene, hexane, m/p xylene, butane, and 2-butanone. The wind information such as the wind directions in the monitoring areas was collected and used to verify that the VOCs observed in these figures were emitted from the industrial complexes of concern. While most of the VOC concentrations detected in the locations downwind were relatively higher than those of the upwind site, it was expected that the VOCs detected in these figures were mostly sourced from the industrial complexes of interest. For few exceptions, additional emission sources near the sites, interferences by the sea and land breeze, or analytical uncertainty at low VOC concentrations are possible explanations [4].

The VOC concentration distributions at different altitudes were considered as the respective chemical fingerprints or concentration profiles. The profiles at different altitudes were compared for discussion of the altitude effect on the VOC distribution and transport. Although toluene consistently dominated at different altitudes through the seasons, the VOC concentration profile at ground level was different from those at higher altitudes, indicating potential vertical transports of the VOCs or different sources responsible for the VOCs present at different altitudes. Table 1 shows the PCA result, which predicted the important VOC speies at ground level and three different altitudes at the monitoring sites in three seasons. Three major principal components that accounted for larger variances of the observed VOC concentrations was listed in the table. The numbers in the parentheses behind the VOC species indicate the rotated factor loadings that describe the influences of this VOC species on each principal component.

Table 1. Principal components of the VOCs observed at different altitudes above the monitoring sites [4]

Altitude	1st principal component (PC) 1	2nd PC	3rd PC
Ground level	Cyclohexane (0.73) Ethylbenzene (0.90) m/p-Xylene (0.90) o-Xylene (0.81) 1-Ethyl-3-Methylbenzene (0.99) 1-Ethyl-4-Methylbenzene (0.99) 1,2,4-Trimethylbenzene (0.99)	Ethane, Ethene, Ethylene (0.97) Isobutane (0.97) 2-Methylbutane (0.98)	1-Butene (0.87) Styrene (0.93)
% of variance	37.5	20.1	14.4
Cumulative variance	37.5	57.6	72.0
100 m	m/p-Xylene (0.84) o-Xylene (0.89) 1-Ethyl-3-Methylbenzene (0.94) 1,2,4-Trimethylbenzene (0.92)	Isobutane (0.81) 1-Butene (0.92) Dichloromethane (0.86)	n-Hexane (0.79) Toluene (0.85)
% of variance	21.5	15.9	15.9
Cumulative variance	21.5	37.4	53.3
300 m	m/p-Xylene (0.76) o-Xylene (0.90) 1-Ehtyl-3-Methylbenzene (0.90) 1,2,4-Trimethylbenzene (0.87)	Isobutane (0.86) 1-Butene (0.76) 2-Methylbutane (0.75) n-Butane (0.76)	Toluene (0.83) 1-Ethyl-4-Methylbenzene (0.78) Dichloromethane (0.92)
% of variance	18.0	17.0	10.1
Cumulative variance	18.0	35.0	45.1
500 m	Toluene (0.69) Isobutane (0.82) 2-Methylbutane (0.87) n-Butane (0.95) Propylene (0.86) 1-Butene (0.86)	o-Xylene (0.90) 1-Ethyl-3-Methylbenzene (0.98) 1-Ethyl-4-Methylbenzene (0.99) 1,2,4-Triethylbenzene (0.89)	Ethane, Ethene, Ethylene (0.76) Sulfur dioxide (0.83)
% of variance	24.8	10.8	10.7
Cumulative variance	24.8	35.6	46.3

In the table, the distributions of the principal components appeared to be different amongst the ground level and three altitudes, with respect to the percentages of the variance accounted for by the three principal components and the species included in each principal component. The VOCs at ground level were heavily loaded by the three principal components. 72.0% of the original VOC data variance could be described by these VOC species in three principal components. As the altitude increased, the contributions of three principal components decreased, somehow suggesting that the grouping amongst the VOCs with similar characteristics and/or behaviors became less obvious. A stronger effect of atmospheric transport and mixing at high altitudes could be one explanation. By comparing the VOC species included in these principal components at ground level and high altitudes, the 1st principal component at ground level mainly included aromatic compounds. Cyclohexane was the exception, as it is an aliphatic compound but with a very similar chemical structure. In the result at 500 m above the ground level, the 1st principal component mainly included saturated and unsaturated aliphatic hydrocarbons, such as alkanes, alkenes, and alkynes. These comparisons suggested that the sources mainly responsible for the VOC pollutions at different altitudes could be different and the molecular characteristcis affected their dispersion, transportation, and mixing depth in the atmoshere. For instance, flare stacks with heights ranging from 100 to 300 m in the examined areas could one important source for the aliphatic hydrocarbons with low molecular weights observed at high altitudes. Traffic was more likely to be associated with aromatic compounds, which were observed at ground level or lower altitudes and have more complex structures and higher molecualr weights.

In Table 1, mono-aromatic hydrocarbons (e.g., ethylbenzene or xylenes) in the 1st principal component at the ground level accounted for 37.5% of the original data variance. These compounds were reported to be more relevant to vehicular activities [20]. As mono-aromatic hydrocarbons were still the important compoenents in the 1st principal component at 100 and 300 m above the ground level, the VOCs in the 1st principal component for the data at 500 m above the ground changed. Most of the compounds in the 1st principal component of the data at 500 m above the gorund level were alkanes and alkenes, as these compounds are known to be emitted from both vehicular and petrochemical industrial sources. The physicochemical characteristics of these VOCs could be another point of view for discussion of the possible VOC sources. The dominance of alkanes at high altitudes were possbily attributable to their relatively long lifetimes, enhancing their accumulation at higher altitudes in the atmosphere. However, it is more difficult for alkenes to transport far from their sources given their shorter lifetimes in the atmosphere. Given that local sources are more important for alkenes than for alkanes, and therefore, the alkanes and alkenes in the 1st principal component at 500 m above the monitoring sites were possbily sourced from the local industrial sources, such as those flare stacks with heights.

Figure 4 is the 3-D loading plots by the cluster analysis to illustrate if similar behaviors existed amongst the VOCs detected at different altitudes. The VOCs observed at ground level were clustered into five distinct groups. It included the aromatics with high loading values on the 1st principal component (Group 1), aliphatic hydrocarbons (mostly saturated alkanes) (Group 2), unsaturated aliphatic hydrocarbons (both alkenes) (Group 3), cyclohexane (Group 4), and the other compounds with low loading values on all 3 principal components (Group 5). The dominant VOCs were aromatic compounds and those with high molecular weights. The grouping phenomenon with respect to their physicochemical characteristics became less significant as the altitude increased. Group 1 (Ethylbenzene and m/o/p-xylene) and 2 (1-Ethyl-3-Methyl-Benzene and 1,2,4-Methylbenzene) were clustered separately. A number of saturated and unsaturated hydrocarbons were clustered in Group 3. n-Hexane and toluene together represent Group 4. The grouping was more unobvious at 300 m above the ground level. The VOCs with similar chemical structures clustered with each other again at 500 m above the ground level. It is worth noting that the group size was smaller as well, similar to the result for the ground level data, indicating a strong "inherent" characteristic in each VOC group at ground level and 500 m above.

The discussion above suggested the VOCs concentration profiles could be different between the ground level and higher altitudes. While aromatic VOCs represent the substantial species accounting for the maximum variances of the data observed at ground level, saturated and unsaturated aliphatic compounds were more associated with the VOC pollutions at high altitude. By identifying the importance of mono-aromatic hydrocarbons at ground level, traffic was obviously an important source for mititgating the VOC contaminations near ground level or at low altitudes in the atmopshere. As the heights of flare stacks in the examined areas mostly ranging from 100 to 300 m, both saturated and unsaturated hydrocarbons responsible for the VOCs at high altitudes in the atmosphere were expected to be emitted from the local industrial activites. The long-range transport of these VOCs at high altitudes were not likely to occur given the relatively short lifetimes of selected VOCs such as alkenes. The PCA and cluster analysis both illustrated a transition trend: clear grouping for the data at ground level and 500 m and obscure differentiation among the groups at 100 and 300 m above the ground level. While many proceeding studies focused on the horizontal distributions of VOC pollutions in the atmopshere, this section attempted to recommend that the influence of altitude on the VOC distribution in the atmosphere is important and not negligible. Moreover, this altitude effect is affected by the locations of emission sources and physicochemical properties of VOCs such as their molecular characteristics and atmospheric lifetime times.

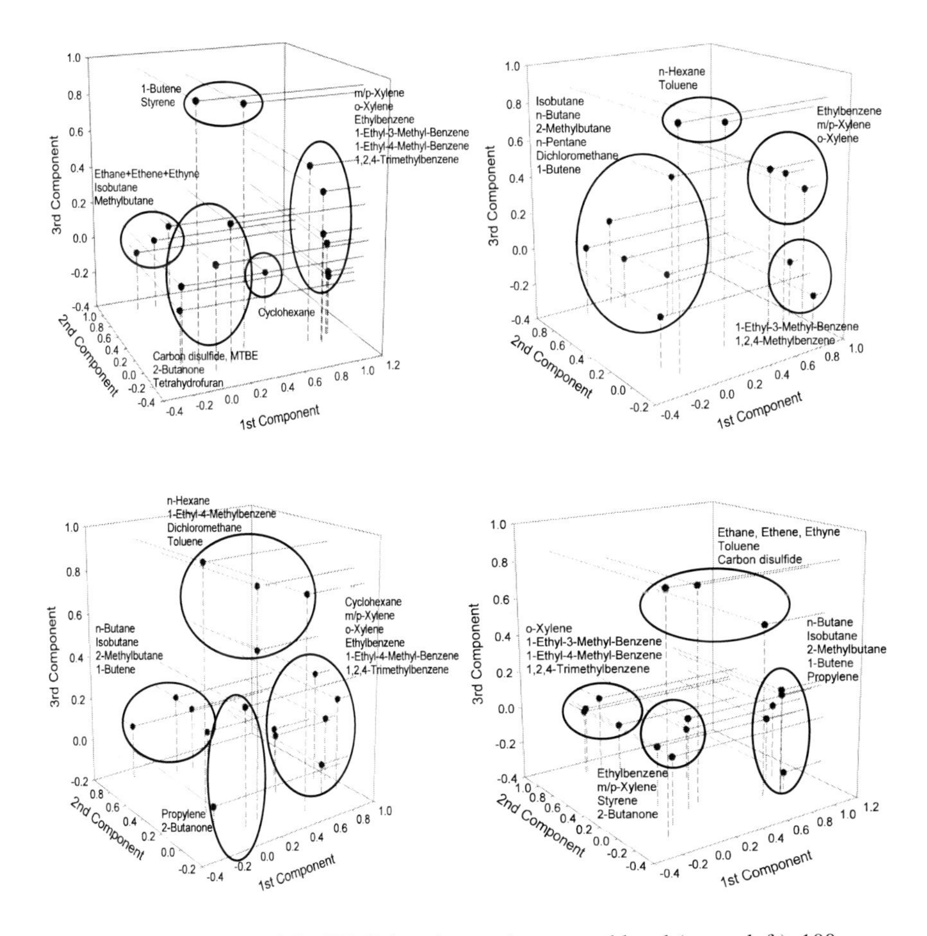

Figure 4. Loading plot of the VOC data detected at ground level (upper left), 100 m (upper right), 300 m (lower left), and 500 m (lower right) above the ground from the PCA and cluster analysis [4].

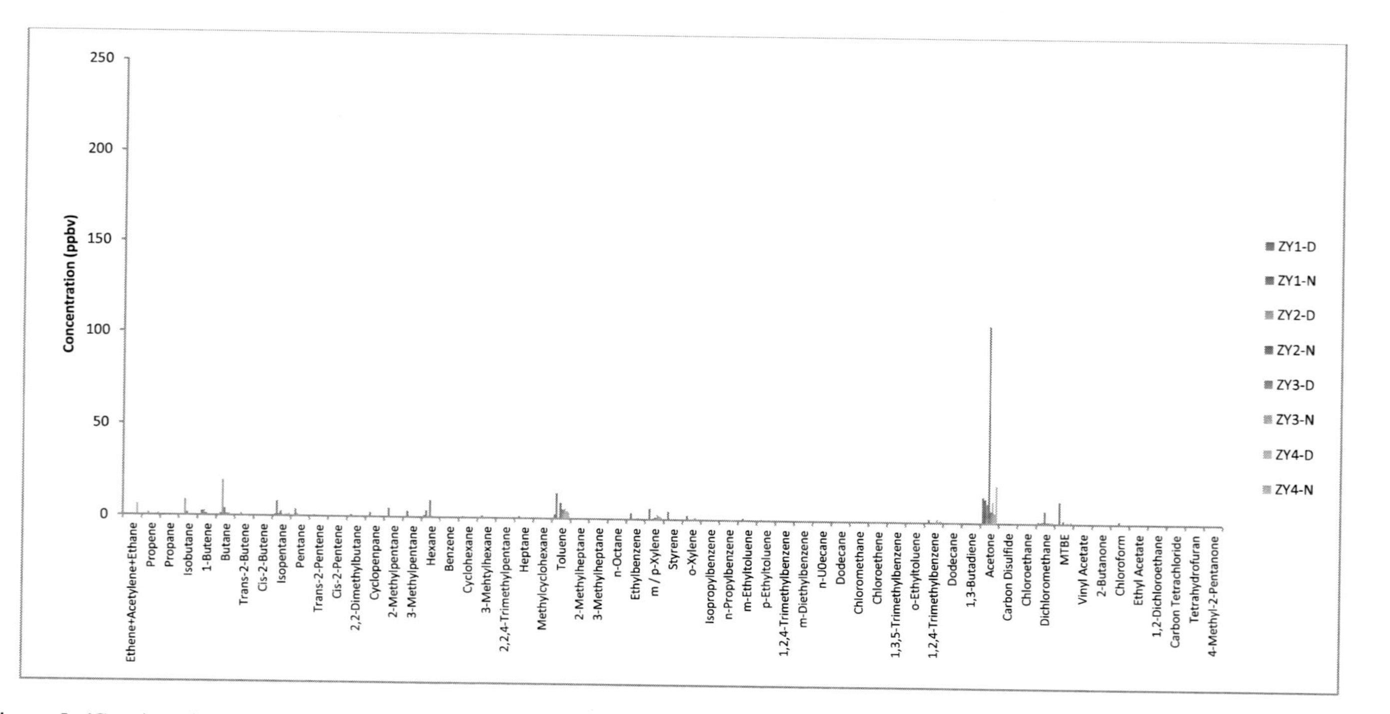

Figure 5. (Continued).

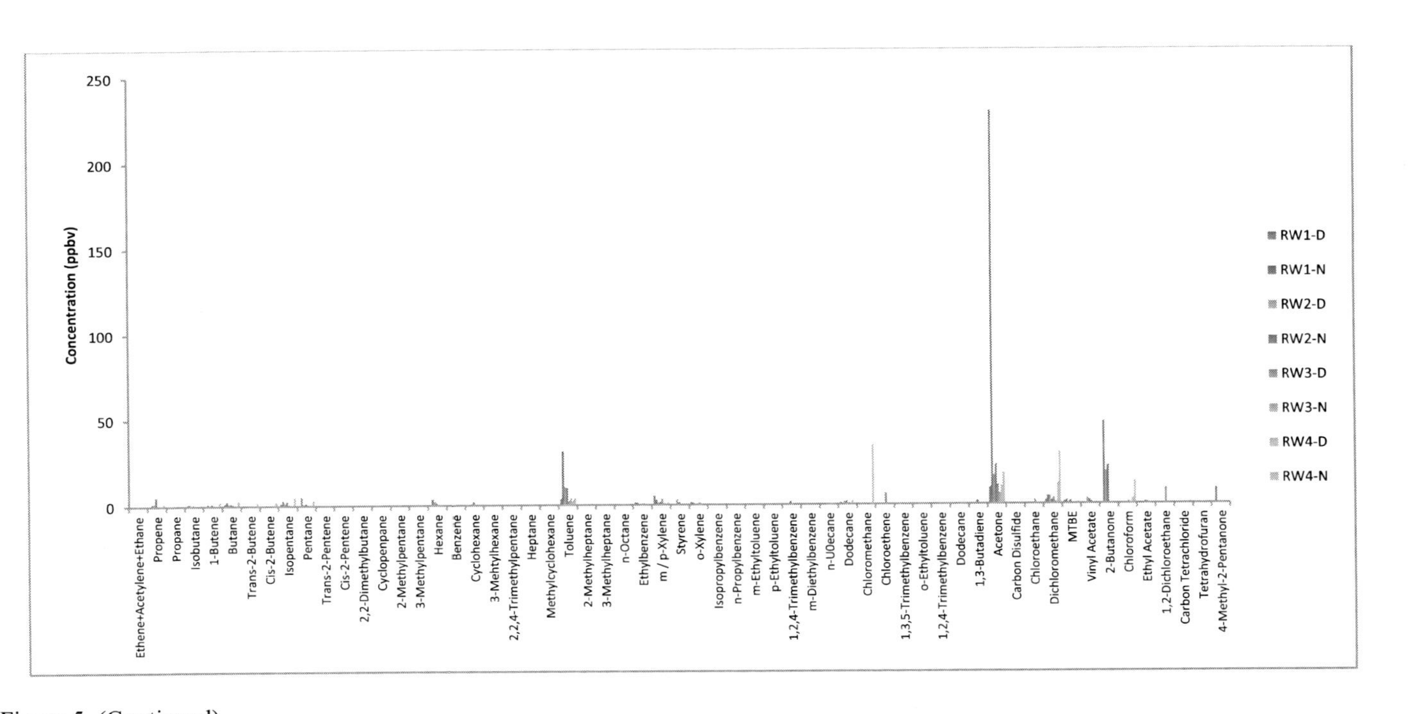

Figure 5. (Continued).

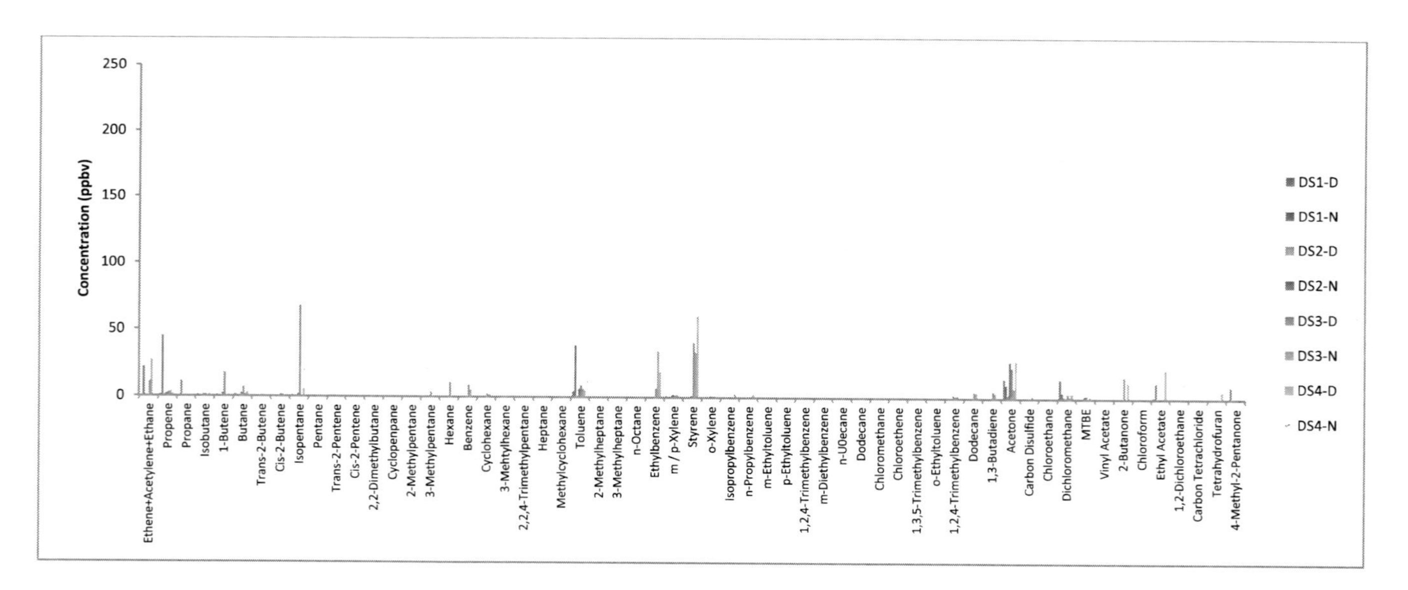

Figure 5. VOC concentration profiles observed at three monitoring sites in the dry season. D and N denote the concentrations observed during the daytime and nighttime, respectively [5].

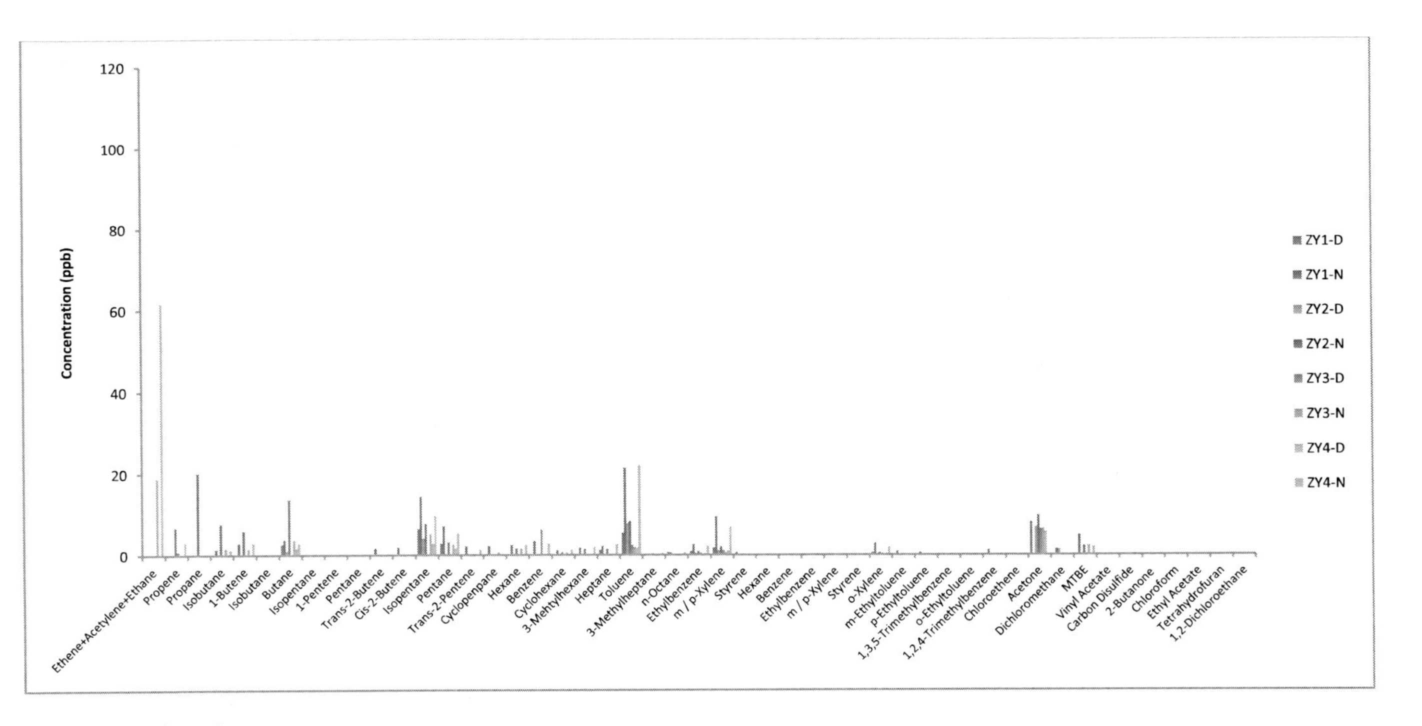

Figure 6. (Continued).

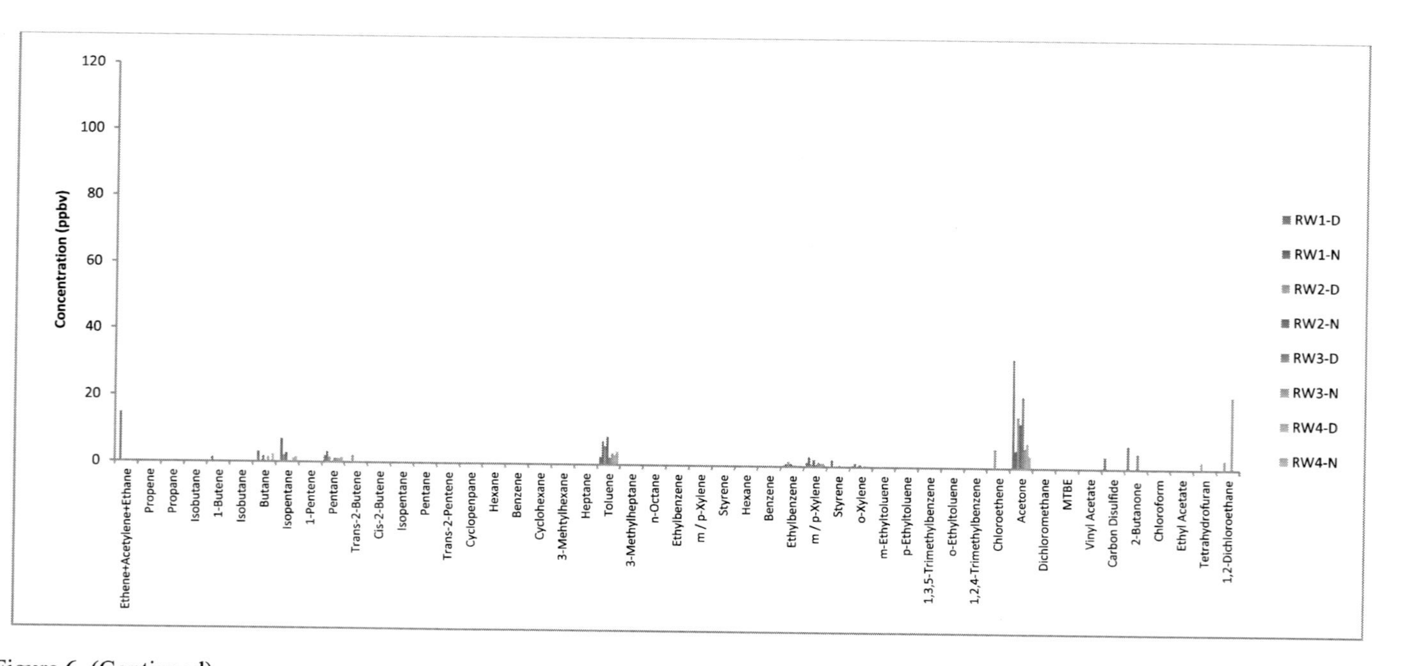

Figure 6. (Continued).

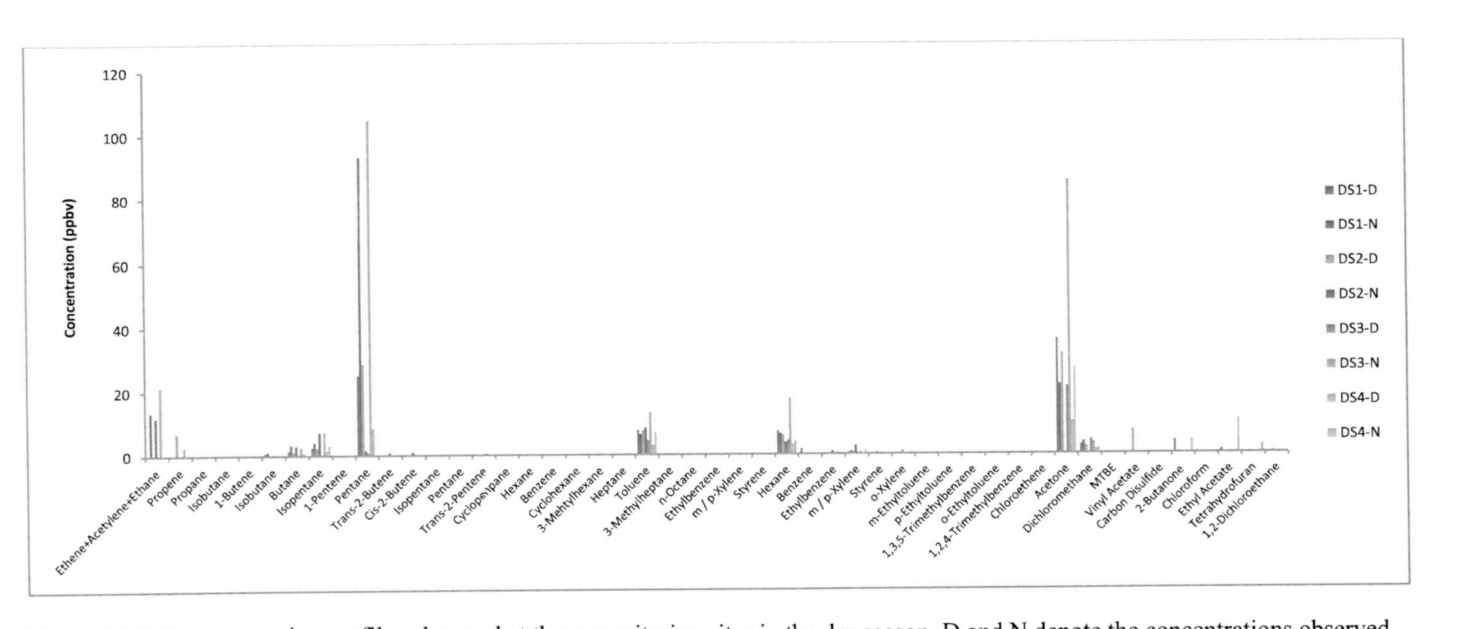

Figure 6. VOC concentration profiles observed at three monitoring sites in the dry season. D and N denote the concentrations observed during the daytime and nighttime, respectively [5].

The Effects of Diurnal Temperature and Seasonal Humidity Variations by Tropical Savanna Climate

Besides the altitude effect, the VOC pollution in Kaohsiung City is also affected by the the Tropical Savanna Climmate in this area [5]. Liu et al. analyzed the VOC concentrations in the atmosphere at three moniotring sites (ZY, RW, and DS sites) in Kaohsiung City during the daytime and nighttime in the dry and wet seasons, as shown in Figures 5 and 6. The figures only illustrate the VOCs observed at appreciable levels. Similarly, acetone and toluene had the relatively higher concentrations in the dry season, as butane, 2-butanone, propene, isopentane, and styrene occurred with more distinguishable concentrations. In the wet season, the higher humidity levels might reduce the concentrations of acetone and toluene by inhibiting the emissions and transports of these compounds in the atmosphere. The concentrations of a number of VOCs observed in the dry seasons were also lower or became negligible in the wet season. Some VOCs were more important in the wet season, such as dichloroethane, propane, butane, pentane, isopentane, and hexane. Interestingly, all these VOCs belong to the saturated hydrocarbons with relatively low molecular weights. As mentioned in the previous section, alkanes are more likely accumulated in the atmosphere due to their longer lifetimes in the atmosphere, and possibly, also more resistant to the influence of humidity in the wet season.

Figures 5 and 6 also show the influence of diurnal temperature variation on the VOC concentration distributions. The VOCs with high concentrations were mostly observed at night. The influence of this diurnal temperature variation on the VOC concentration distributions was expected to be assoicated with the changes of temperature or extent of sunlight in the atmosphere. Different physicochemical characteristics affected the capabilities of different VOCs to handle these environmental variations. However, the effect of diurnal temperature variation on the VOC concentrations appeared to interact with the effect of seasonal humidity variation, as the later could be more important. For example, a number of VOCs were present at the monitoring sites during both daytime and nighttime, whereas the number and concentrations of many VOCs detected in the wet season significantly reduced possibly due to the humidity effect. Figure 7 plots the loading values of the VOCs in three principal components of the data from the dry and wet seasons to characterize the humidity effect on the VOC behaviors. The VOCs observed at four monitoring sites in the dry season were clustered into 7, 4, and 4 distinct groups. The cluster number in the ZY area was more than the numbers in the other two monitoring areas, indicating that the VOC concentration distribution in this area was more complex. In the wet season, the VOCs observed in the RW and DS areas were more clearly grouped and the spread of each cluster appeared to be smaller, implying more pronounced inherent characteristics of distributions in each VOC cluster. The cluster number in the ZY area was also reduced from seven to four in the wet season. One possible explanation was that certain VOCs were more resistant to the effect of humidity variation, possibly attributed to their different physicochemical properties, and therefore, became more dominant amongst the VOCs in the wet season. For instance, the VOCs listed in the 1st principal component in the wet season were mostly aromatic and cycloaliphatic compounds.

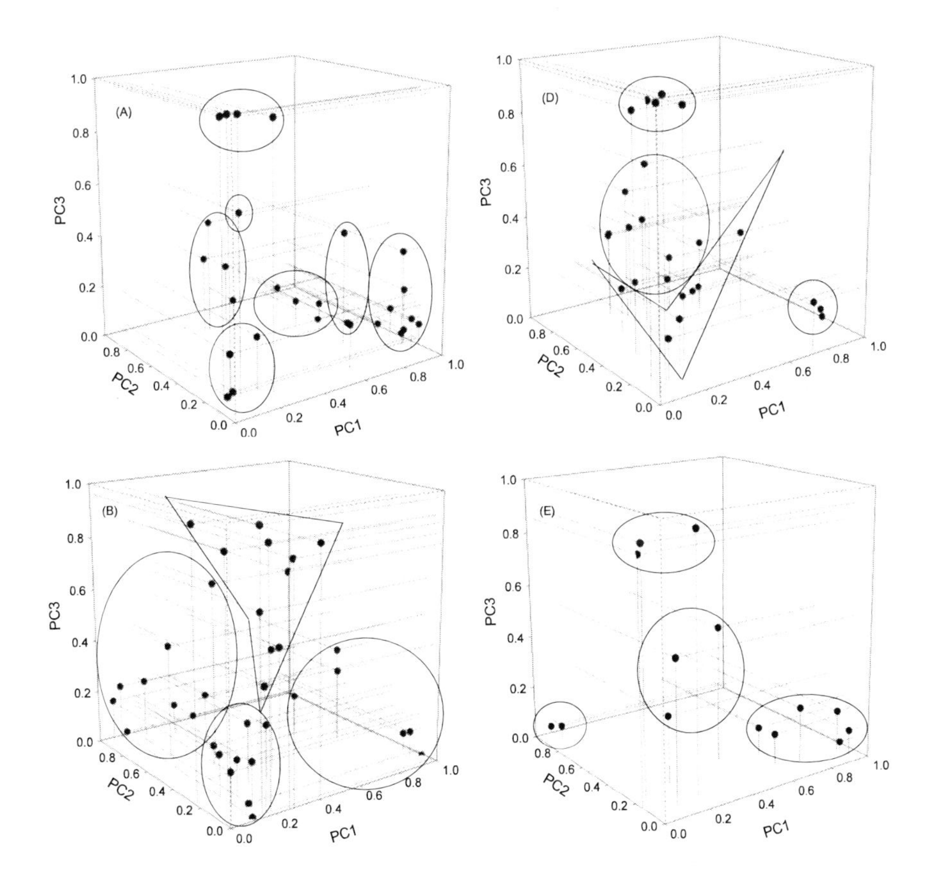

Figure 7. (Continued).

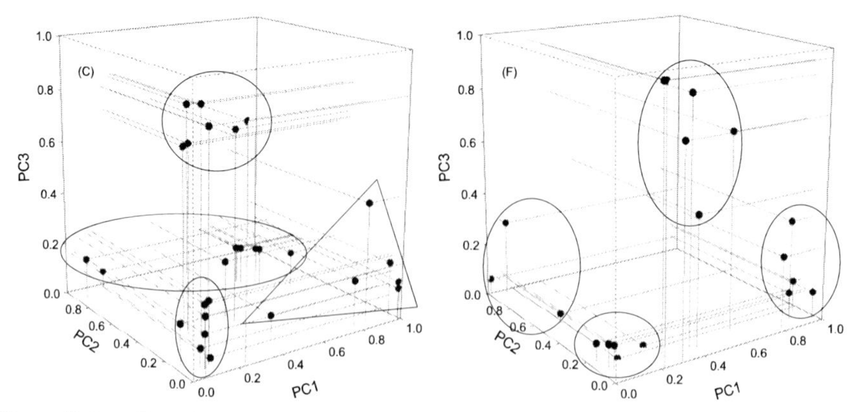

Figure 7. Loading plots of the VOC data observed at four sites: (A)ZY, (B)RW, and (C)DS in the dry season and (D)RW, (E)ZY, and (F)DS in the wet season [5].

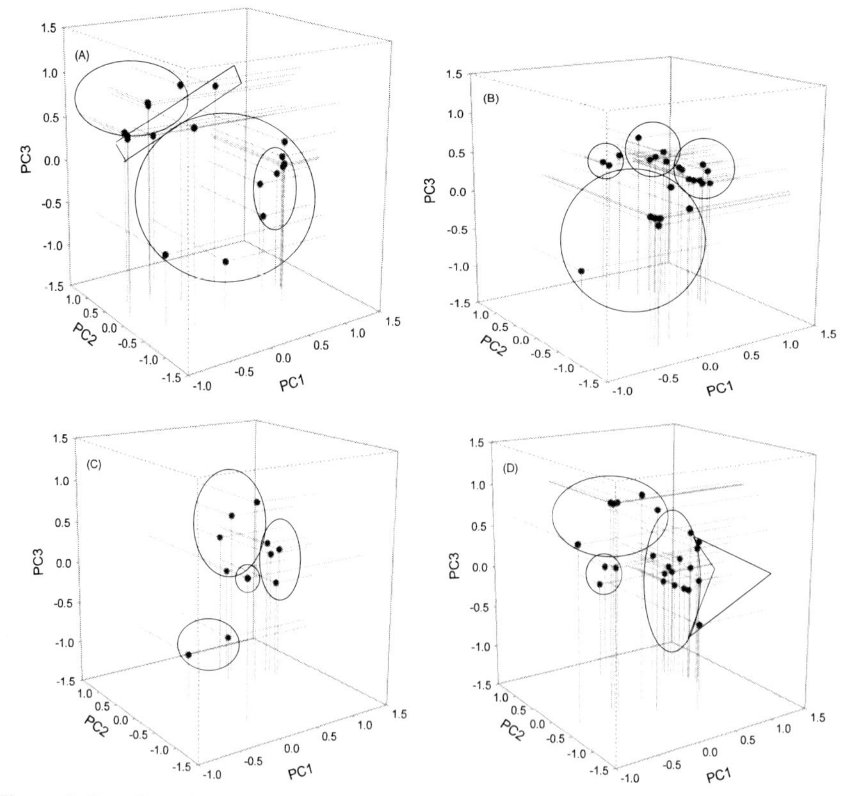

Figure 8. Loading plots of the VOC data observed at the ZY area during the (A) daytime and (B) nighttime in the dry season and during (C) daytime and (D) nighttime in the wet season [5].

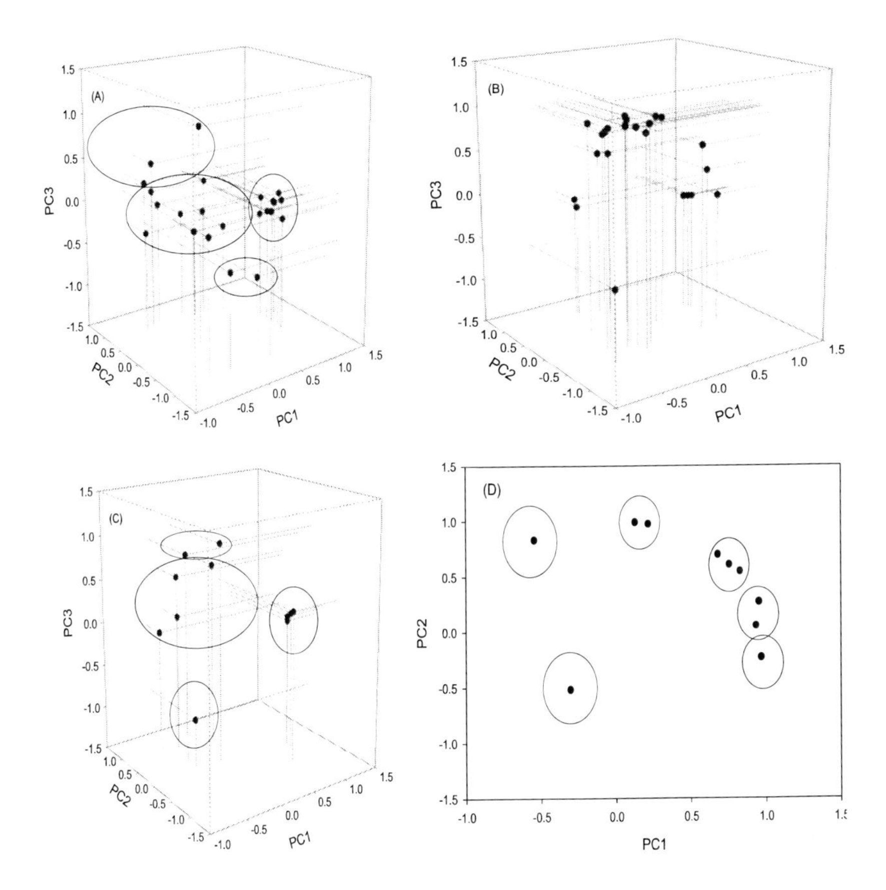

Figure 9. Loading plots of VOC data observed in the RW area during the (A) daytime and (B) nighttime in the dry season and during the (C) daytime and (D) nighttime in the wet season [5].

Figures 8 through 10 show the loading values of the VOCs in three principal components during the daytime and nighttime to illustrate the influence of diurnal temperature variation on the VOC concentration distribution. No matter in either dry or wet season, the grouping was more definite and precise during the daytime, particularly in the ZY and RW areas, whereas the spreads of the groups became larger during the nighttime. The result indicate a less pronounced inherent characteristic in each VOC group during the nighttime. It is possible that the VOCs dominant during the daytime were more persistent given their stronger photochemical resistances or longer atmospheric lifetimes. At night, other VOCs more vulnerable to sunlight or high temperatures occurred and affected the results of the PCA and cluster analysis. On the other hand, the influence of high temperature and sunlight was not present at night. More VOCs could be present in the monitoring data, diluting the contributions of dominant VOCs observed during the daytime to the full concentration distributions at night.

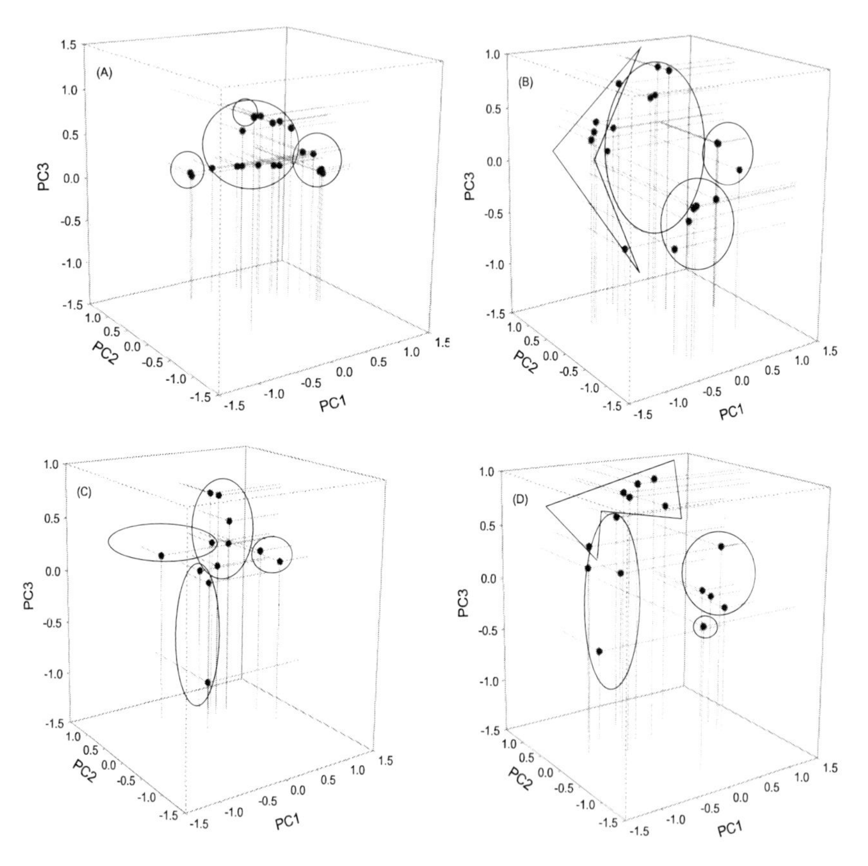

Figure 10. Loading plots of VOC data observed in the DS area during the (A) daytime and (B) nighttime in the dry season and during (C) daytime and (D) nighttime in the wet season [5].

Liu et al. reported the influences of diurnal temperature and seasonal humidity variations caused by a tropical savanna climate on the VOC concentration distributions in the atmosphere [5]. The stronger photochemical reactions and elevating humidity seemed to reduce the VOC concentrations and inhibit the distributions during the daytime and in the wet season. Compared to the influence of diurnal temperature variation on the VOC concentration distribution, the effect of seasonal humidity variation seemed to be more important. In the wet season, aromatic hydrocarbons with more complex structures and high molecular weights and selected aliphatics accounted for the maximum variances of the VOC data observed. For the effect of diurnal temperature variation, the VOCs with stronger photochemical resistances and longer atmoephric lifetimes were the dominant species observed during the daytime, whereas additional VOCs more vulnerable to sunlight or high temperatures also occurred at night. These findings demonstrated the non-negligible effects of diurnal temperature and seasonal humidity variations by a tropical savanna climate on the VOC emissions and distributions. The variations of humidity and temperature (or intensive sunlight) which could happen in many regions of the world potentially augmented the characteristics and impacts of VOC pollutions from anthropogenic sources on the environment and public health.

Aromatic VOCs in the WWTP and the Effects of Activated Sludge and Aeration

Besides industrial complexes, WWTPs represent another important VOC emission source. Chen et al. investigated the VOC emissios in treatment of wastewater by conducting bench-scale simulation experiments and moniotring the VOC concentrations in a full-scale WWTP [6-9]. Figure 11 shows the fugacity of benzene among the air, water, and sludge phases in a bioreactor experiment that simulated the activated sludge in a WWTP. When a fugacity ratio in any phases was less or larger than 1, the compound was less or more saturated in that phase than its reference phase, and thus, was more or less likely to transfer from the water to the phase of interest, respectively. In Figure 11, the transfer of benzene was from water to air phases before aeration begun in the first 2 hours of the experiments due to the high benzene concentration in the untreated wastewater. The transferring potential significantly dropped as aeration started from the 3rd hour. Even though aeration was expected to enhance the benzene emission from the water to air phases, it also reduced the benzene concentration in the water phase and indirectly reduced the concentratin in the air phase by mixing and dilution with fresh air. With the continuous aeration, the transfer of benzene was reversed and from air to water as the benzene concentration in the water phase became low enough in the 5th hour in the experiment. The water was not aerated in the last 2 hours, and benzene was transferred from water to air again. It is worth noting that benzene was transferred from the sludge to water phases through the experiment, suggesting that the direction of this transfer was not affected by aeration, and more importantly, benzene were not effectively removed by activated sludge in this biological treatment process.

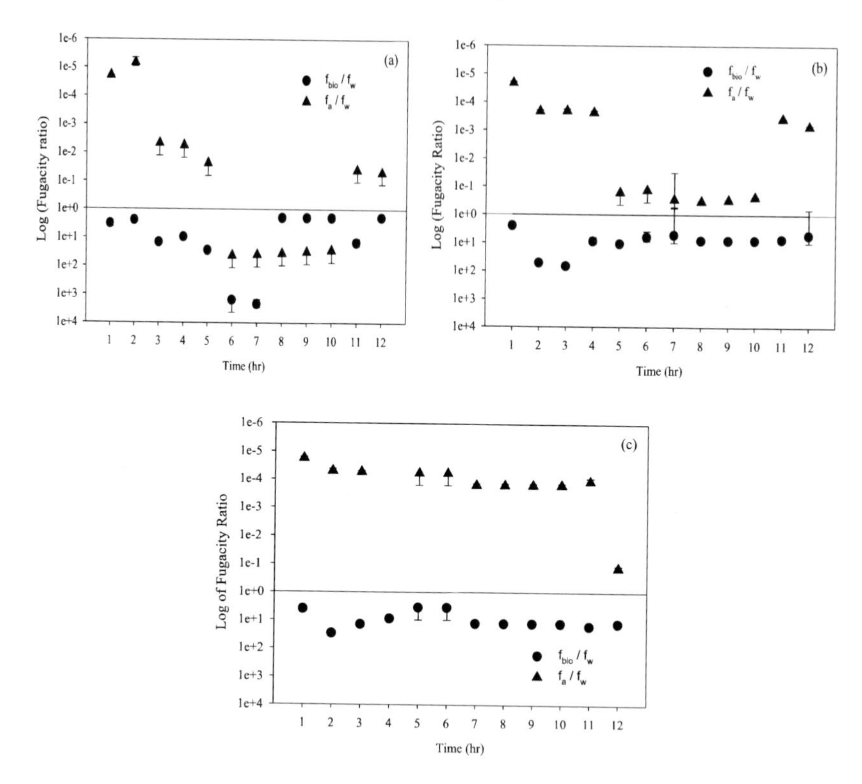

Figure 11. Fugacity of (a) benzene, (b) toluene, and (c) xylenes in the phases of air, water, and activated sludge in the simulation experiments, assuming the fugacity in the water phase was equal to 1 [8].

Figure 11 also showed the results of toluene and xylene in the air and sludge phases of the simulation experiments. All three compounds were transferred from sludge to water and from water to air phases with strong potentials in the primary sedimentation. From the 3rd hour, the aeration reduced the transfers of these VOCs, and the extent of the effect decreased in the order of benzene > toluene > xylenes. Aeration might help reudce the water-phase concentrations of these VOCs more significantly than their air-phase concentrations, diminishing the concentration gradients between the water and air phases and lowering the transfering potentials. The directions or extents of transferrig between the water and sludge phases were not substantially different amongst benzene, toluene, and xylenes. It followed in the results that these aromatic VOCs were not effectively adsorbed or removed by the activated sludge in the experiments, and aeration did not help improve the situations.

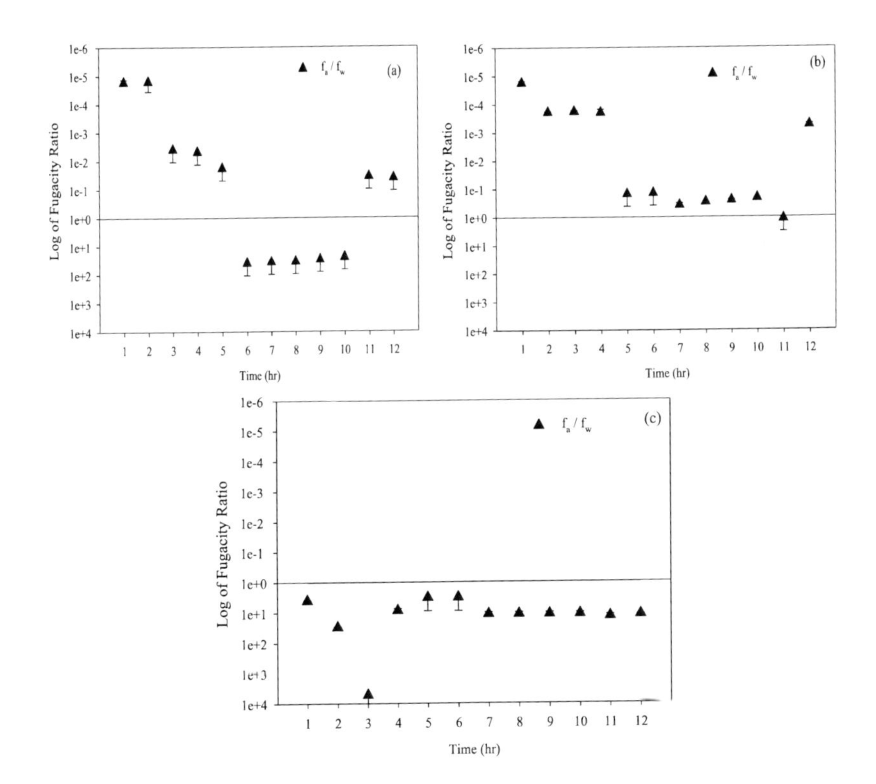

Figure 12. Fugacity of (a) benzene, (b) toluene, and (c) xylenes in the phases of air, water, and activated sludge in the simulation experiments without activated sludge, assuming the chemical activities was equal to 1 (the straight line) [8].

Figure 12 illustrates the results when the simulation experiments were conducted with aeration without addition of activated sludge into the bioreactor. In the comparison between Figures 11 and 12, the effect of adding activated sludge was negligible on the transfers of benzene and toluene from the water to air phases. However, the presence of activated sludge in the bioreactor significantly changed the transfer of xylenes between the water and activated sludge phases. With a higher K_{OW} value of xylenes, the activated sludge provided a pathway of biosorption and/or biodegradation for xylenes to reduce its water-phase concentration and potentially reversed the direction of the transfer. The individual effect of aeration was studied by comparing the Figures 11 and 13.

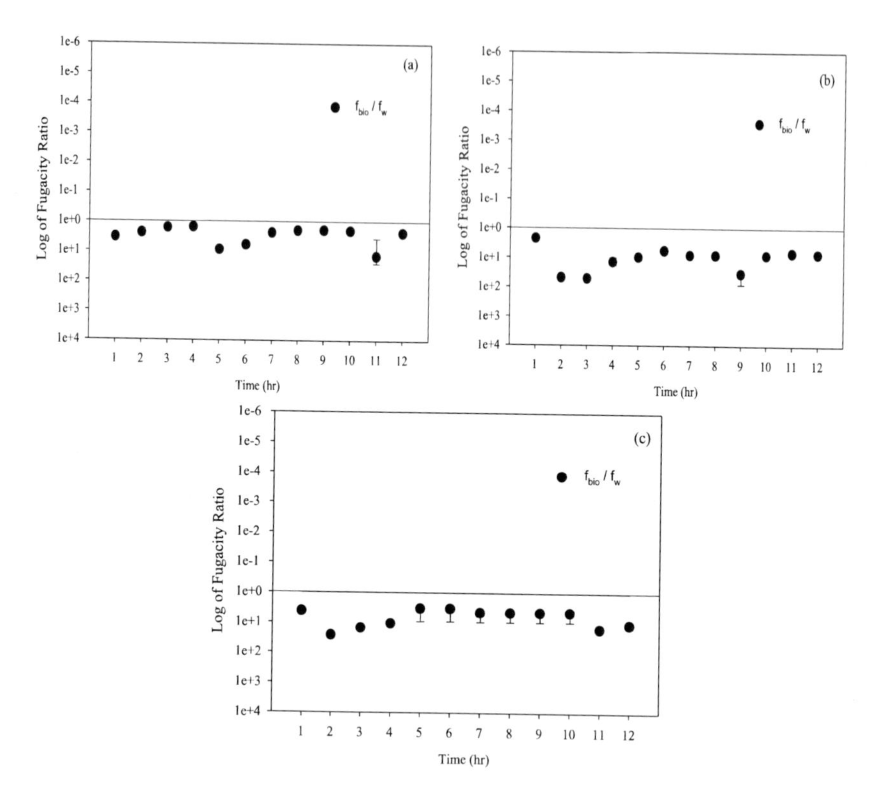

Figure 13. Fugacity of (a) benzene, (b) toluene, and (c) xylenes in the phases of air, water, and activated sludge in the simulation experiments without aeration, assuming the chemical activities was equal to 1 (the straight line) [8].

Different from the effect of adding activated sludge, the aeration was more effective to change the transferring potential of benzene from the activated sludge to water phases. The effect was rather limited for toluene and xylenes. The influence of aeration was stronger on benzene, potentially due to its lower K_{OW} value, decreasing its concentration gradient and transferring potential between the water and sludge phase. Overall, the presence of activated sludge without aeration did not help remove the VOCs in the water phase. It is worth noting that the exact effect of aeration was difficult to predict with it simultaneously enhancing volatilization and biosorption/biodegradation by increasing the emission rates and the activity of sludge in the reactor, respectively.

Table 2. Summary of the influences of aeration on the VOC concentrations in different phases[8]

	Physicochemical properties of the substance			
	K_H		K_{OW}	
	Low[1,2]	High[1,3]	Low[1,4]	High[1,5]
Phase	Effect of aeration on reducing the phase concentration of the substance			
Air	Weak	Strong	-	-
Water	Strong	Weak	Strong	Weak
Activated Sludge	-	-	Weak	Strong
	Positive effect of aeration on the direction of transferring potential			
	From air to water	From water to air	From activated sludge to water	From water to activated sludge

[1] With only three compounds being investigated in this study, the effect of aeration on the fates of the VOCs was described qualitatively rather than quantitatively. However, the numbers of these parameters of the VOCs investigated in this study were given here for readers' references.

[2] Such as benzene in this study, with a Henry's law coefficient of 5.5 x 10^{-3} atm-m^3/mol.

[3] Such as xylenes in this study, with a Henry's law coefficients raging from 5.18 x 10^{-3} to 7.18 x 10^{-3} atm-m^3/mol.

[4] Such as benzene in this study, with an octanol-water partition coefficient of 2.13.

[5] Such as toluene and xylenes in this study, with octanol-water partition coefficients ranging from 2.72 to 3.2.

While the transfer of a VOC between the air and water phases were affected by its K_{aw} value, the K_{OW} value and biodegradability are important to determine its potentials for biosorption and/or biodegradation and thus its sludge-phase concentration. The individual and co-effects of these factors including aeration, activated sludge, and the physicochemical properties of VOCs are complex and affect the fates of VOCs among the air, water, and sludge phases in a biological treatment process. Overall, the transfers of the VOCs between the water and activated sludge phases were limited, and volatilization from the water to air phases was an important pathway for the VOC removal in the wastewater. Activated sludge in the bioreactor helped reduce the transfers of aromatic VOCs with higher hydrophobicity from the water to air phases. The effect of aeration on the transfers of VOCs was more complex, as a number of factors could be involved, as shown in Table 2. For the VOCs with large K_H and small K_{OW} values, aeration more substantially reduced the concentration in the air and water phases, enhancing its volatilization and diminishing the biosorption and/or biodegradation in the biological treatment process such as activated sludge.

Chlorinated VOCs in the WWTP and the Effects of Activated Sludge and Aeration

The mass fractions of four chlorinated VOCs including chloroform, carbon tetrachloride, trichloroethylene (TCE), and tetrachloroethylene (PERC) in three phases of the activated sludge simulation experiments were shown in Figure 14. The chlorinated VOCs were mostly dissolved in the procsses. Even though the addition of sludge which was supposed to enhance the VOC treatment via biological treatment, most VOCs were either present in the water phase or vaporized. It indicates that besides aromatics, chlorinated VOCs were not effectively treated but emitted into the atmosphere in WWTPs. For each chlorinated VOC, less than 1% of the mass was subject to biological treatment in the experiments. The partitions of these chlorinated VOCs became more favorable to the air phase after the aeration started, and reached apparent plateaus in the 2nd half of the experiment. Aeration is originally expected to support the activated sludge in a biological treatment process so the VOCs in the water phase can be more effectively and efficiently removed. Nevertheless, the finding here indicates a negative impact of aeration on the performance of biologically treating the chlorinated VOCs, as the aeration actually promoted the VOC emissions into the atmosphere in the treatment process.

Figure 14 also implies that the chemical structure of a VOC affected its mass distributions. In the result, in comparison with the unsaturated TCE and PERC, chloroform and carbon tetrachloride belong to saturated hydrocarbons and were more easily removed by volatilization or biological sorption and degradation. In addition, more chlorine in the strucutres of saturated chloroform and carbon tetrachloride inhibited their volatilization and limited the biological treatability. However, the results were opposite for TCE and PERC. As halogen substitution is supposed to increase the hydrophobicity of a

VOC by reducing its polarity, the results pointed out that this influence of halogen substitution was different when different VOCs were considered, implying the complex behaviors of VOCs in the biological treatment process such as activated sludge given a number of VOCs with diverse physicochemical properties present in wastewaters.

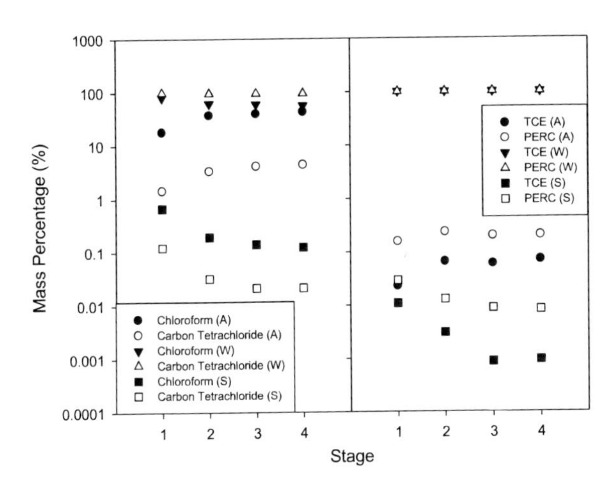

Figure 14. Mass distributions of four chlorinated VOCs in the simulated activated sludge experiments. TCE and PERC denote trichloroethylene and tetrachloroethylene, respectively. The 4 stages represent the pre-sedimentation without aeration for the first 2 hours, followed by the forepart (from the 3rd to 6th hour) and rear part (from the 7th to 10th hour) of aerobic biological treatment process, and post-sedimentation without aeration in the last 2 hours [6].

Figure 15 compares the partitions of chlorinated VOCs among three phases in the simulation experiments with the known partition coefficients. Besides the K_{aw} and sludge-water partitioning coefficients (K_{sludge}) values, the K_p values were provided in the study and calculated by using the concept of linar free energy relationship to predict the sorption onto the surface of sewage sludge ($\log K_p = 0.58 \times \log K_{ow} + 1.14$). In the results, the K_{sludge} values were slightly lower than the Kp values because of the lipid and protein fractions of the sludge sampled (5% liqid and 25% protein). Next, except for carbon tetrachloride, the sludge-water partition fractions of these chlorinated VOCs were mostly larger than their air-water partition fractions. A higher sludge-water partition fraction suggest the potential for treating a VOC in biological wastewater treatment processes. More importantly, the sludge-water or air-water partition fractions of four VOCs estimated in the experiments were all below the reported partition coefficients. It followed that the VOCs were still present in the water phase instead of being evaporized or biologically removed. It must be noted that the partition fractions of the VOCs were not significantly varied as the eperimental situation was changed. The physicochemical property of a VOC such as its volatility or polarity could be more important factors to determine its fate through the wastewater treatment processes.

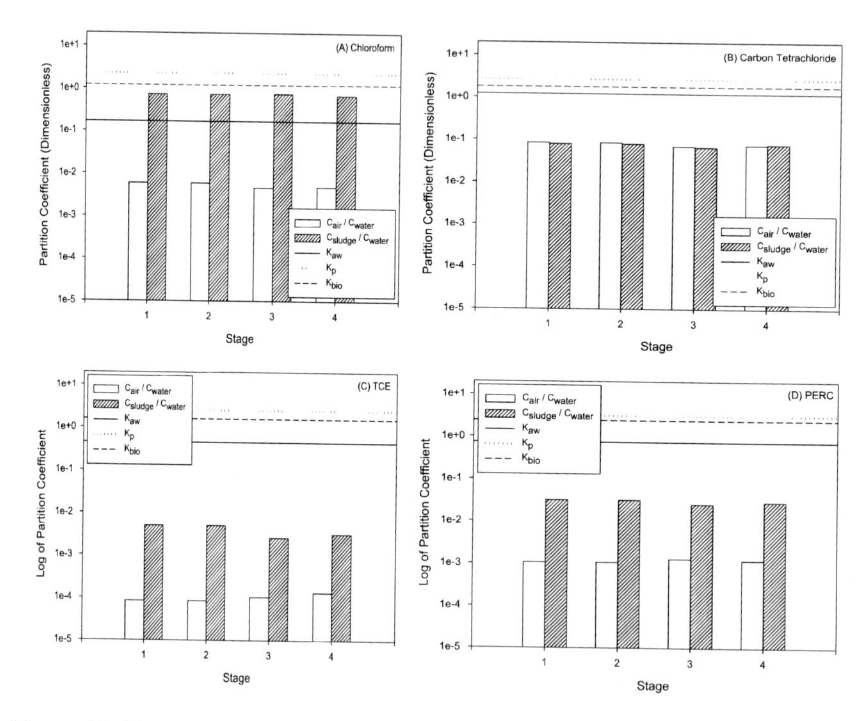

Figure 15. Air-water and sludge-water partition fractions of chlorinated VOCs through four wastewater treatment stages and their reported K_{aw}, K_p, and K_{sludge} values. TCE and PERC denote trichloroethylene and tetrachloroethylene, respectively [6].

The fugacity of four chlorinated VOCs in three phases in the activated sludge simulation expriments was shown in Figure 16. It was noted that four chlorinated VOCs were transferred from the water to air or sludge phases with strong potentials. Aerating enhanced the volatilization but limitedly improved the biological sorption and degradation. Although four VOCs in the water phase were continuously vaporized and bioloigcally removed, the extents of their transferring potentials were different among the species. Volatilization was more important for the VOC with less chlorine in the structure such as chloroform and TCE, even though the K_H values of the saturated carbon tetrachloride and unsaturated PERC were larger given more chlorine in their chemical structures. The extent of biological sorption and degradation for removal decreased in the order of TCE > PERC ≥ carbon tetrachloride > chloroform. The dimensionless Kow value diminished in the order of PERC > carbon tetrachloride > TCE > chloroform. The different orders in this comparison might indicate the co-effects of other factors such as their biodegradability.

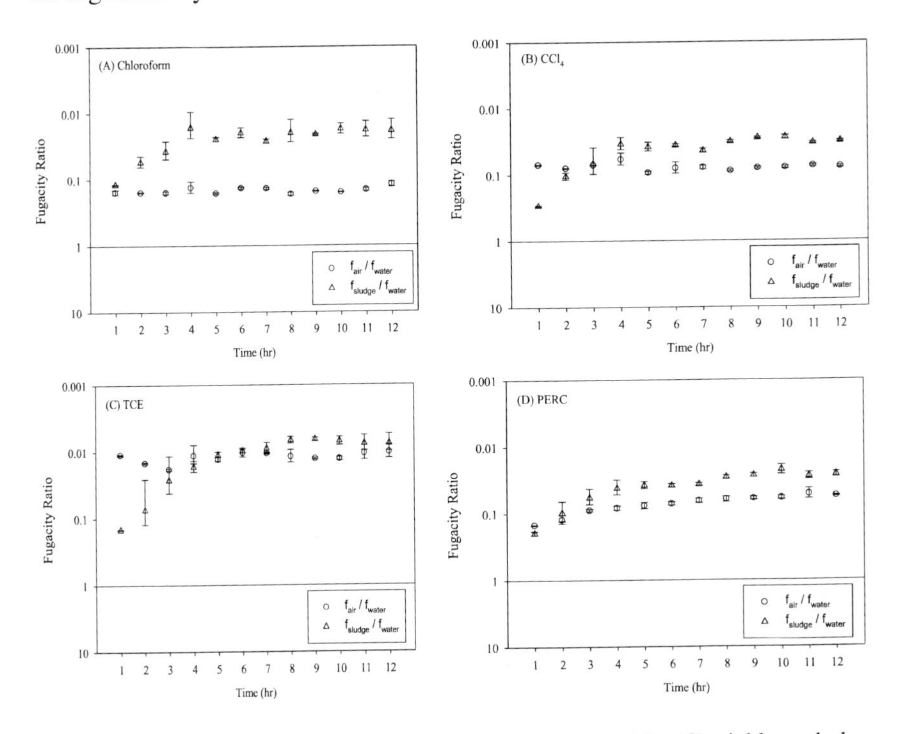

Figure 16. Fugacity of (A) chloroform; (B) carbon tetrachloride; (C) trichloroethylene (TCE); and (D) tetrachloroethylene (PERC) in the phases of air, water, and sludge in the activated sludge simulation experiments, assuming that the fugacity of the compounds in the water phase equal to one. The four stages of the experiment include the pre-sedimentation without aeration for the first 2 hours, followed by the forepart (from the 3rd to 6th hour) and rear part (from the 7th to 10th hour) of aerobic biological treatment process, and post-sedimentation without aeration in the last 2 hours [6].

Figure 17A investigates the individual effect of aeration by preparing the same experiments without the presence of activated sludge. The influence of aeration appeared to be limited on volatilization of chloroform and carbon tetrachloride. Aeration was more influential for the unsaturated VOCs, as aeration appreciably enhanced and inhibited the potentials of PERC and TCE for partitioning into the air phase. The K_H values of TCE and PERC were similar (0.020 and 0.018 atm-m^3/ml, respectively). Figure 17B illustrates the individual effect of sludge addition on the fates of chlorinated VOCs without aeration. All four VOCs were increasingly transferred to the sludge phase for biological sorption and/or degradation. The case of TCE was more noticeable, possibly attributable to its biodegradability. Regardless of saturated or unsaturated VOCs, slightly stronger potentials for biological treatment occurred for the VOCs with less chlorine such as chloroform and TCE. By comparing these figures, it is noteworthy that the volatilization of TCE was weakened with aeration and the presence of sludge. TCE was apt to be present in the dissolved form or removed by biological sorption or degradation in the activated sludge.

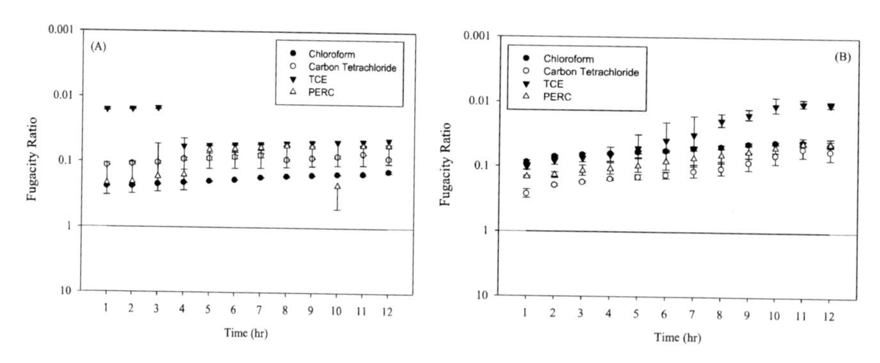

Figure 17. Fugacity of chloroform, carbon tetrachloride, trichloroethylene (TCE), and tetrachloroethylene (PERC) in the phases of air, water, and sludge in the activated sludge experiments (A) without sludge addition and (B) without aeration, assuming that the fugacity of the compounds in the water phase equal to one. The four stages of the experiment include the pre-sedimentation without aeration for the first 2 hours, followed by the forepart (from the 3rd to 6th hour) and rear part (from the 7th to 10th hour) of aerobic biological treatment process, and post-sedimentation without aeration in the last 2 hours [6].

Mass and Concentration Distributions

Yang et al. compared the mass distributions of three aromatic and chlorinated hydrocarbons in three phases in the activated sludge process, as shown in Figure 18 [7]. The air phase was the main compartment in which aromatic hydrocarbons were mostly present. Aromatic VOCs were continuously volatilized from the water phase at the beginning of the experiment. Although the K_H values of xylenes were relatively higher amongst these VOCs, it was more distributed in the water phase, possibly explained by its higher K_{OW} that also affected the distributions in three phases in a biological treatment process. The transferring potentials of these aromatics were enhanced as the water was aerated at the 2nd hour of the experiment. Of the three aromatics, xylenes was the one more easily removed from the water phase. Given their higher K_{OW} values by additional methyl functional groups in the chemical structures, toluene and xylenes were more efficiently removed via biosorption or biodegradation in the sludge phase. Aeration and its physicochemical properties (e.g., K_H and K_{OW}) played two important roles to detemine the treatability of a VOC via volatilization or biosorption/ biodegradation in wastewater treatment processes.

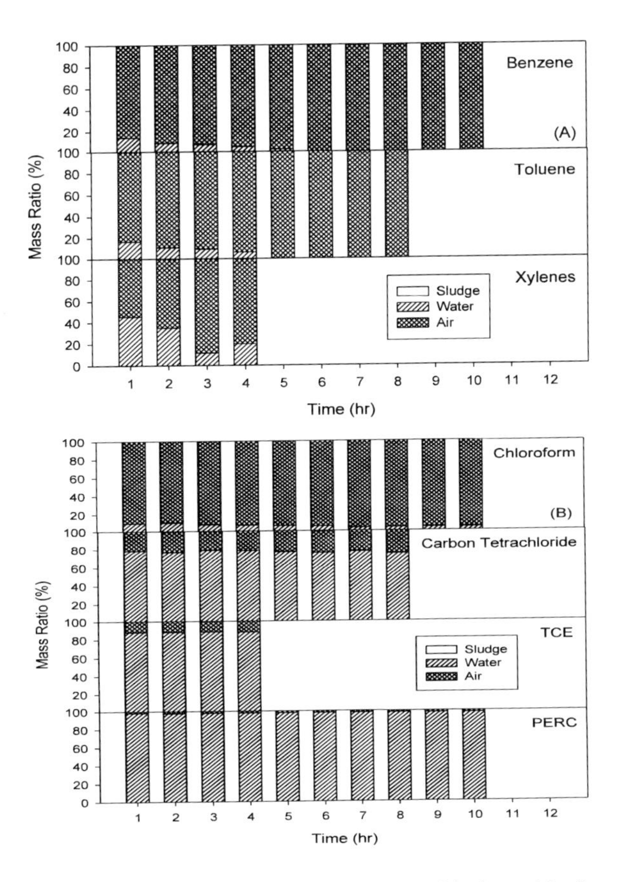

Figure 18. Mass distributions of (A) aromatic and (B) chlorinated hydrocarbons in the air, water, and sludge phases in the simulated biological reactors [7].

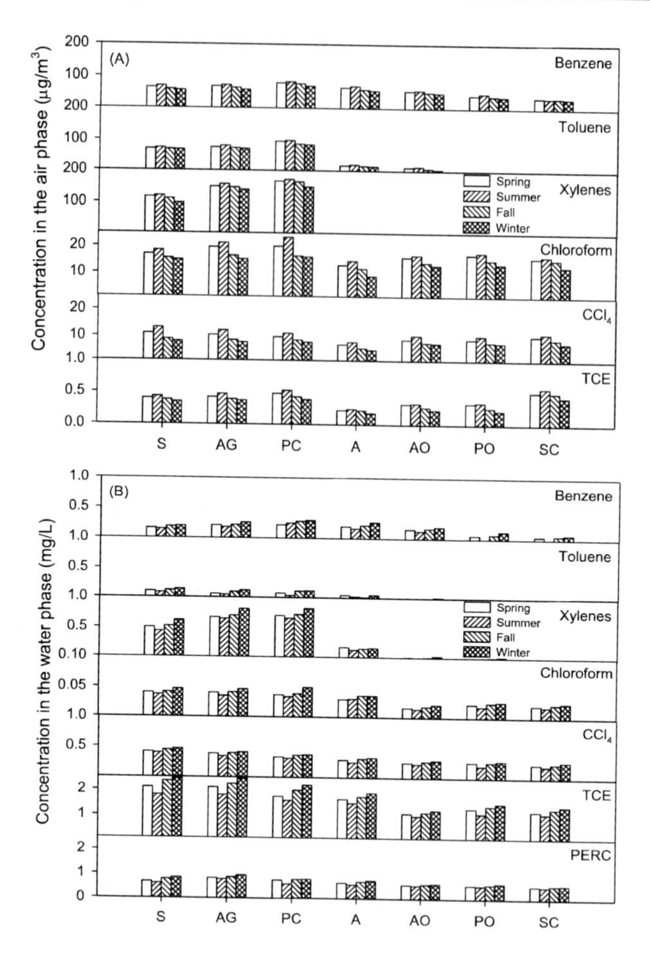

Figure 19. Concentrations of aromatic and chlorinated hydrocarbons in the (A) air and (B) water phases in a full-scale WWTP (S, AG, PC, A, AO, PO, and SC denote the screen, aerated grit chamber, primary clarifier, anaerobic tank, anterior oxic tank, posterior oxic tank, and secondary clarifier, respectively) [7].

Emission Rates

Given the observations in the previous section, the emission rates these aromatic and chlorinated VOCs in the full-scale wastewater treatment processes were calculated and compared for discussion [9], as present in Figures 20 through 24.

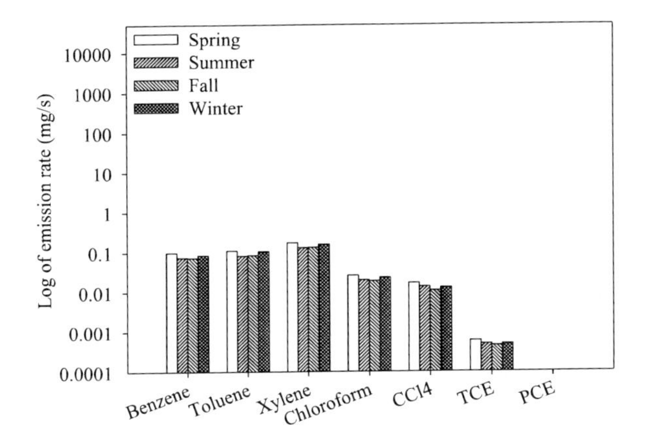

Figure 20. Emission rates of the VOCs of interest in a full-scale bar screen process [9].

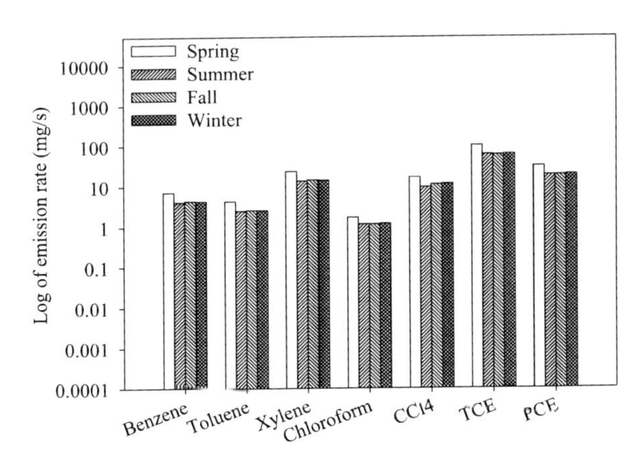

Figure 21. Emission rates of the VOCs of interest in a full-scale primary sedimentation process [9].

Of different treatment processes, the highest emission rates occurred in the aerobic sludge treatment process, whereas the lowest emission rates were observed in the bar screen. The emission rates decreased in the order of the aerobic sludge treatment process > primary and secondary sedimentation basins > anaerobic sludge treatment process > bar screen. The explanations for different emission rates in different processes were possibly associated with whether the process was a close or open system, surface areas, fetch-to-depth ratios, and the water-phase VOC concentrations. For example, the higher emission rates observed in the primary sedimentation basins and anaerobic sludge treatment process might be explained by their larger surface areas (1963 and 1000 m^2, respectively) and higher water-phase concentrations.

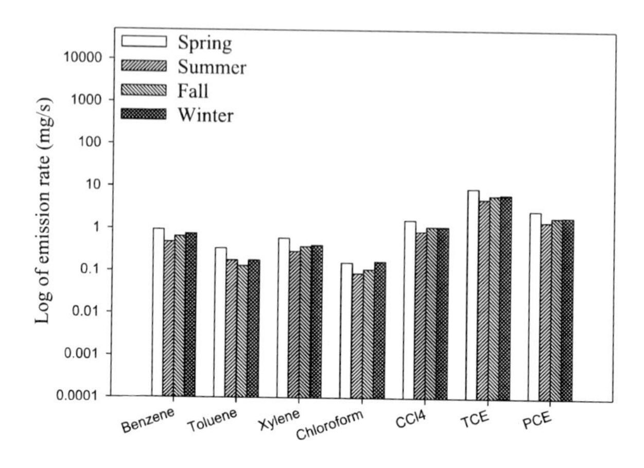

Figure 22. Emission rates of the VOCs of interest in a full-scale anaerobic sludge treatment process [9].

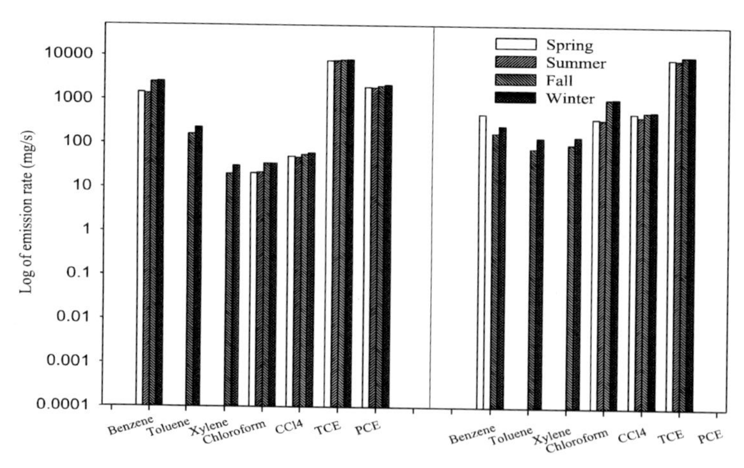

Figure 23. Emission rates of the VOCs of interest in the forepart (left) and rear part (right) of a full-scale aerobic sludge treatment process [9].

Of different VOC species, the emission rates of the aromatic VOCs were higher than those of chlorinated hydrocarbons in the bar screen. The aromatics VOCs had relatively higher emission rates in the bar screen, which was assumed to be a close system. However, in the primary and secondary sedimentation basins and anaerobic sludge treatment process, which were considered as open systems, the chlorinated VOCs had higher emission rates. Benzene and TCE were two VOCs with higher emission rates in the full-scale treatment processes. The factors including the physicochemical properties of the VOCs such as volatility, biodegradability, and water-phase concentrations simultaneously affected the emission rates of these VOCs in different processes. It was noted in the results that the decreasing emission rates among the aromatic VOCs could be associated with the additional number of methylene groups in the molecular structure. However, the finding was different for the chlorinated VOCs, as the chlorine number and chemical bonding type in the strucutres were more influential to for their emisison rates. Their limited treability via biological processes also incurred their higher VOC emission rate.

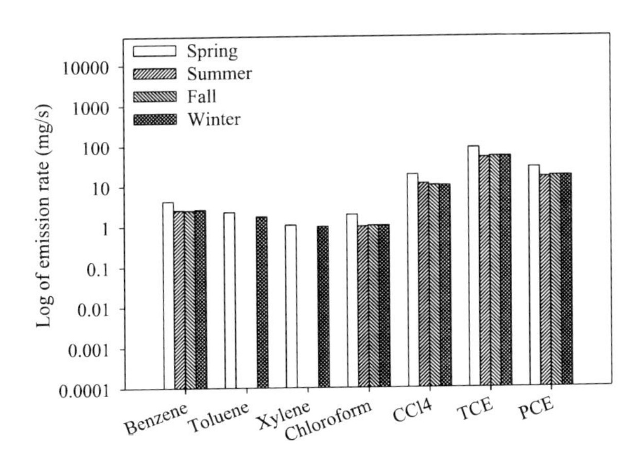

Figure 24. Emission rates of the VOCs of interest in a full-scale secondary sedimentation basin [9].

Of different seasons, the higher emission rates of the VOCs occurred in spring. The VOC emissions were expected to be affected by the ambient air temperature and wind speed. However, the air or water temperature variation was limited during the monitoring periods in the study, limiting the effect of temperature on the emission rates. The wind speed was one more important explanation for the high emission rates in spring. In the primary and secondary sedimentation basins and anaerobic sludge treatment process, which were considered as open system, the high wind speeds in spring enhanced the VOC emission rates. Although the bar screen was a cose sytem, the wind speed flowing into the process was also significantly higher in spring. Different patterns observed in the forepart and rear part of the aerobic sludge treatment process were explained by the influence of aeration, which could be more important than that of seasonal temperature variation. Simply speaking, different emission rates observed among the seasons were more associated with the wind speed variation, instead of the air or water temperature variations.

Table 3. Excess lifetime cancer risks for a full-scale WWTP's workers exposed to the carcinogenic VOCs [9]

	Benzene				Chloroform			
	Spring	Summer	Fall	Winter	Spring	Summer	Fall	Winter
Bar screen	9.02E-08	9.73E-08	8.32E-08	7.89E-08	4.13E-09	4.50E-09	3.77E-09	3.62E-09
PS[1]	1.11E-07	1.17E-07	1.09E-07	9.87E-08	4.86E-09	5.69E-09	3.99E-09	3.91E-09
Anaerobic S[1]	9.16E-08	9.87E-08	8.32E-08	7.89E-08	3.16E-09	3.55E-09	2.84E-09	2.16E-09
Aerobic S (F)[1]	7.75E-08	8.04E-08	7.19E-08	6.91E-08	3.89E-09	4.08E-09	3.40E-09	3.16E-09
Aerobic S (R)[1]	5.92E-08	6.77E-08	5.64E-08	5.36E-08	4.13E-09	4.30E-09	3.65E-09	3.28E-09
SS[1]	5.08E-08	4.79E-08	4.93E-08	4.65E-08	3.84E-09	3.99E-09	3.69E-09	3.04E-09
	Carbon Tetrachloride				Trichloroethylene			
	Spring	Summer	Fall	Winter	Spring	Summer	Fall	Winter
Bar screen	2.25E-08	2.70E-08	1.82E-08	1.67E-08	3.89E-11	4.18E-11	3.69E-11	3.40E-11
PS[1]	2.00E-07	2.29E-08	1.78E-08	1.61E-08	4.67E-11	5.15E-11	4.18E-11	3.79E-11
Anaerobic S[1]	1.41E-07	1.59E-07	1.14E-08	1.00E-08	2.14E-11	2.33E-11	2.14E-11	1.75E-11
Aerobic S (F)[1]	1.80E-07	2.10E-08	1.57E-08	1.51E-08	3.11E-11	3.21E-11	2.63E-11	2.24E-11
Aerobic S (R)[1]	1.82E-07	2.08E-08	1.59E-08	1.55E-08	3.21E-11	3.35E-11	2.53E-11	2.14E-11
SS[1]	2.08E-08	2.27E-08	1.82E-08	1.53E-08	4.96E-11	5.54E-11	4.86E-11	4.18E-11

[1] PS, Anaerobic S, Aerobic S (F), Aerobic (R), and SS denote the primary sedimentation basins, anaerobic sludge treatment process, the forepart and rear part of the aerobic sludge treatment process, and secondary sedimentation basins, respectively.

[2] The lifetime cancer risks from the exposure to toluene and xylenes were not predicted since the inhalation slope factors of these two substances are not available.

Table 4. Non-carcinogenic risks for a full-scale WWTP's workers exposed to the aromatic VOCs [9]

	Benzene				Toluene			
	Spring	Summer	Fall	Winter	Spring	Summer	Fall	Winter
Bar screen	1.13	1.20	1.07	0.97	0.26	0.27	0.25	0.24
PS[1]	1.32	1.38	1.28	1.17	0.32	0.33	0.29	0.29
Anaerobic S[1]	1.08	1.17	0.98	0.93	0.07	0.09	0.07	0.07
Aerobic S (F)[1]	0.92	0.95	0.85	0.82	0.05	0.06	0.04	0.03
Aerobic S (R)[1]	0.70	0.80	0.67	0.63	N.A.	N.A.	N.A.	N.A.
SS[1]	0.60	0.57	0.58	0.55	N.A.	N.A.	N.A.	N.A.
	Xylenes							
	Spring	Summer	Fall	Winter				
Bar screen	0.50	0.52	0.49	0.47				
PS[1]	0.56	0.57	0.55	0.50				
Anaerobic S[1]	N.A.	N.A.	N.A.	N.A.				
Aerobic S (F)[1]	N.A.	N.A.	N.A.	N.A.				
Aerobic S (R)[1]	N.A.	N.A.	N.A.	N.A.				
SS[1]	N.A.	N.A.	N.A.	N.A.				

[1] PS, Anaerobic S, Aerobic S (F), Aerobic (R), and SS denote the primary sedimentation basins, anaerobic sludge treatment process, the forepart and rear part of the aerobic sludge treatment process, and secondary sedimentation basins, respectively. N.A. denotes not available because the gas-phase concentration of the compound was not detected.

[2] The lifetime cancer risks from the exposure to toluene and xylenes were not predicted since the inhalation slope factors of these two substances are not available.

Cancer and Non-Cancer Risk Assessments

One of the main concerns for the VOCs in the environment is their potentials to pose adverse health effects to human health. Table 3 lists the excess lifetime cancer risks for the workers in a full-scale WWTP from the exposures to aromatic and chlorinated VOCs in the plant [9]. The non-cancer risk predictions were given in Table 4 and 5 [9]. The primary sedimentation basins exhibited the higher cancer risks. Similar trends were found for the non-carcinogenic risks. Although the differences of adverse health risks among the processes were limited, the emission rates wwere higher in the aerobic sludge treatment process. Through conparison, a process that was not aerated and possesses higher VOC emission rates caused higher cancer and/or non-cancer risks to the workers in the WWTPs. However, when the water was aeraed in the selected processes, the VOC emission rates were enhanced but VOC concentrations in the air phase could be mixed and diluted, resulting in lower health risks in the process such as the aerobic sludge treatment process.

Table 5. Non-carcinogenic risks for a full-scale WWTP's workers exposed to the chlorinated VOCs [9]

	Chloroform				Carbon Tetrachloride			
	Spring	Summer	Fall	Winter	Spring	Summer	Fall	Winter
Bar screen	0.07	0.07	0.06	0.05	0.28	0.33	0.22	0.21
PS[1]	0.07	0.08	0.05	0.05	0.26	0.31	0.22	0.20
Anaerobic S[1]	0.04	0.05	0.04	0.03	0.25	0.28	0.22	0.20
Aerobic S (F)[1]	0.05	0.06	0.05	0.04	0.17	0.20	0.14	0.12
Aerobic S (R)[1]	0.06	0.06	0.05	0.05	0.22	0.26	0.19	0.19
SS[1]	0.05	0.05	0.05	0.04	0.22	0.26	0.20	0.19
	Trichloroethylene				Perchloroethylene			
	Spring	Summer	Fall	Winter	Spring	Summer	Fall	Winter
Bar screen	7.00E-04	7.83E-04	6.50E-04	6.17E-04	N.A.	N.A.	N.A.	N.A.
PS[1]	8.00E-04	8.83E-04	7.17E-04	6.50E-04	N.A.	N.A.	N.A.	N.A.
Anaerobic S[1]	3.67E-04	4.00E-04	3.67E-04	3.00E-04	N.A.	N.A.	N.A.	N.A.
Aerobic S (F)[1]	5.33E-04	5.50E-04	4.50E-04	3.83E-04	N.A.	N.A.	N.A.	N.A.
Aerobic S (R)[1]	5.50E-04	5.75E-04	4.33E-04	3.67E-04	N.A.	N.A.	N.A.	N.A.
SS[1]	8.50E-04	9.50E-04	8.33E-04	7.17E-04	N.A.	N.A.	N.A.	N.A.

[1] PS, Anaerobic S, Aerobic S (F), Aerobic (R), and SS denote the primary sedimentation basins, anaerobic sludge treatment process, the forepart and rear part of the aerobic sludge treatment process, and secondary sedimentation basins, respectively. N.A. denoted not available because the gas-phase concentration of the compound was not detected.

[2] The lifetime cancer risks from the exposure to toluene and xylenes were not predicted since the inhalation slope factors of these two substances are not available.

To compare the risks posed by different VOCs, the cancer risks decreased in the order of benzene ≅ carbon tetrachloride > chloroform > TCE. More importantly, the cancer incidences of four carcinogenic VOCs were below 10^{-5}, indicating that risks of workers by the exposures to the VOCs in the WWTP were moderate. The non-carcer risk for workers in the WWTP decreased in the order of benzene > xylenes > carbon tetrachloride ≅ toluene > chloroform > TCE > PCE. Benzene was the compound of more concern given that its HQ exceeded 1. From the viewpoints of emission rate and health risk, benzene was the species with a higher emission rate and adverse health (both cancer and non-cancer) risks to local workers among the VOCs studied. TCE was the species with high emission rate but low adverse health (both cancer and non-cancer) risks. The toxicity such as the slope factor and reference concentration of a VOCs was important to determine whether a VOC was of essential concern for the health of local workers. The importance of VOCs with respect to their adverse health risks to WWTP workers was not exactly correlated to their emission rates in the treatment processes.

The influnece of seasonal temperature variation was also observed in these tables. Spring and summer were two seasons with relatively higher cancer and the non-cancer risks. Notwithstanding the effect of seasonal temperature variation occurred, the adverse health risk variations among different treatment processes or VOC species were more significant, possibly attributable to the limited changes of the ambient air temperature, water temperature, or wind speed among the seasons. Among different treatment processes, the adverse health risks to workers were relatively higher in the sedimentation process. The VOC emission rates were enhanced by aeration in the aerobic sludge treatment process, but nevertheless, due to the strong mixing and dilution with fresh air by aeration the cancer and non-cancer risks were low. Spring and summer were two seasons with higher VOC emission rates and both cancer and non-cancer risks. However, the variations among four seasons with respect to the adverse health risks posed by these VOCs were limited. Without aeration causing turbulences at the water surface, the health risks of these VOCs were strongly associated with their emission rates, while the findings were different with aeration as the VOC emission rates were enhanced without significantly elevating the associated adver health risks.

Public Health Risks Associated with the VOC Emitted from a WWTP

Besides workers in a WWTP, the health of public, notably those living near the WWTP, may also be adversely affected by the VOC emitted, dispersed, and transported for long distances from the WWTP. Yang et al. used the VOC concentration information given in the previous sections of this chapter and the Industrial Source Complex (ISC) model to predict their transports from the WWTP, followed by calculating the cancer and non-cancer risks to investigate the question [7]. The area of interest was 10 km x 10 km near the WWTP.

The environmental and meteorological information was collected and the WWTP was assumed as the single point source. Figure 25 shows the similar patterns among the concentration distributions of different VOCs during the monitoring periods. The VOCs followed the wind direction toward the southwest, and the transports were inhibited by a mountain located east of the WWTP. However, the ranges of their tranports and distributions varied. The areas affected by the occurrences of chlorinated VOCs were relatively larger that those of aromatic VOCs.

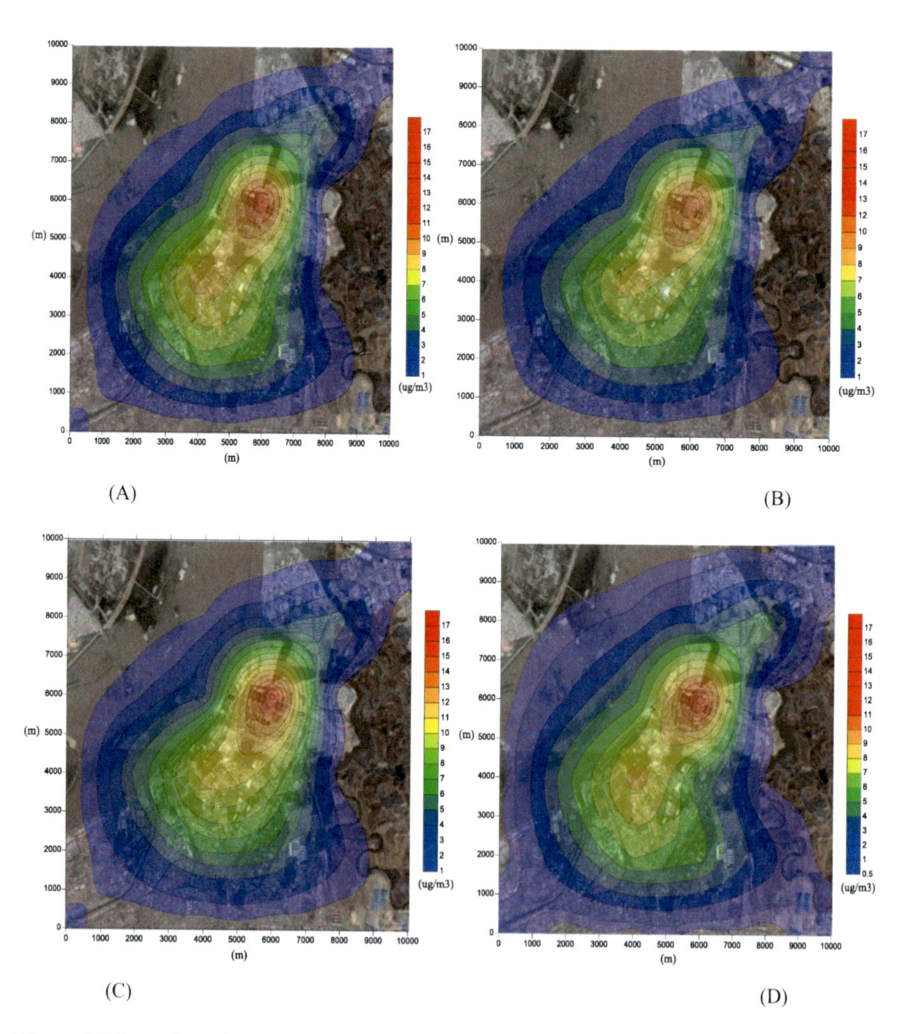

Figure 25 (Continued)

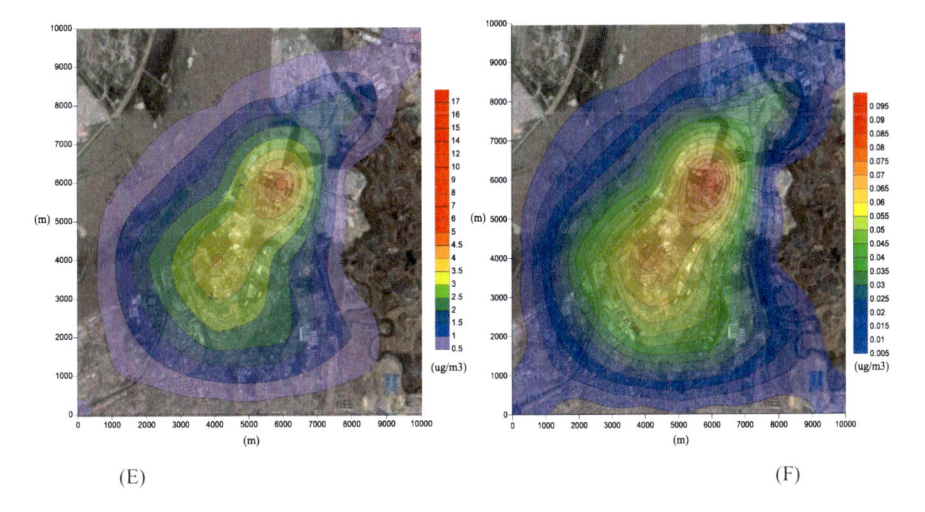

Figure 25. Concentration distributions (μg/m^3) of (A) benzene, (B) toluene, (C) xylenes, (D) chloroform, (E) carbon tetrachloride, and (F) trichloroethylene in the area near a WWTP [7].

The individual and total cancer risks caused by the VOC distributions in Figure 25 were shown in Figure 26 [7]. It was noticed that the populations living in the areas even more than 4 km away from the WWTP were still exposed to a cancer risk above 10^{-6} (a typical regulatory threshold limit). Chloroform was the most important species to cause the risks, whereas the risks posed by TCE were limited. The risks of benzene and carbon tetrachloride were approximately one order of magnitude lower than that of chloroform. This finding followed that the VOCs emitted from a WWTP could be one of the important concerns for the chronic adverse health risks for both the workers in the WWTP and those people living some distances away from the plant. It is worthnoting that an accumulation of the VOCs emitted from the WWTP was found at a location away from the source location given the environmental and metrological conditions, as shown in Figure 26D or 26E. Overall, as the physicochemical property and toxicity of a VOC was more important to determine its risk compared to the effects of variations of seasonal temperature and treatment technology, the adverse health risks posed by the VOCs from a WWTP could mainly resulted from the occurrences of few selected VOCs such as chloroform in this case of the study.

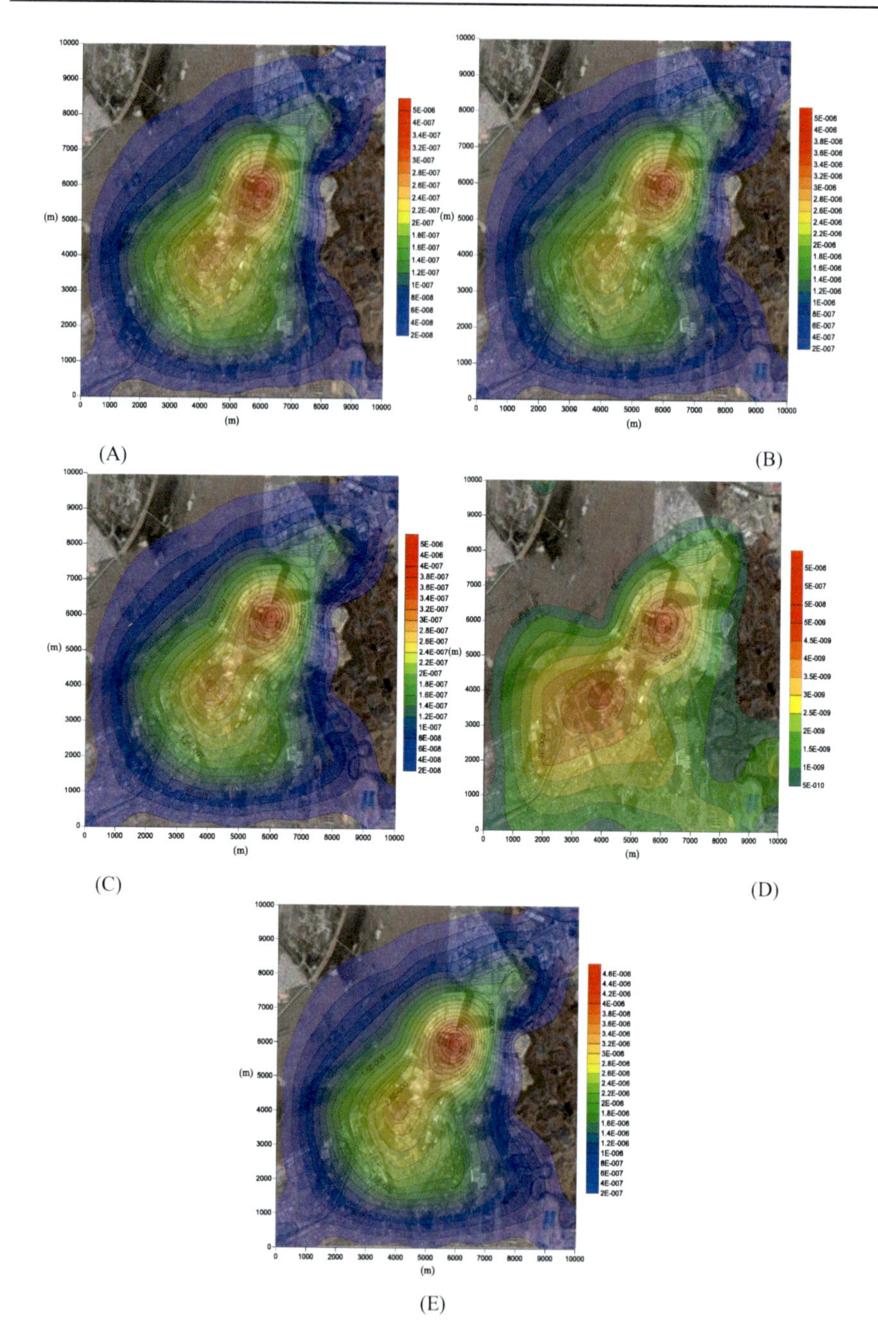

Figure 26. Cancer risk posed by (A) benzene; (B) chloroform; (C) carbon tetrachloride; (D) TCE and (E) total four VOCs emitted from the WWTP of concern [7].

CONCLUSION

This chapter attempted to investigate the VOC pollutions from industrial complexes and WWTPs by starting from summarizing selected articles and discussing the effects of altitude and tropical savanna climate on the VOC distributions in Kaohsiung City of southern Taiwan. Furthermore, additional articles that investigated the VOC emission rates, mass distirbutions among different phases (e.g., air, water, and sludge), and the associated adverse health risks to local workers and populations living near a WWTP were selected for discussion. For the altitude effect, while aromatic VOCs represent the substantial species accounting for the maximum variances of the data observed at ground level, saturated and unsaturated aliphatic compounds were more associated with the VOC pollutions at high altitude. Traffic was obviously an important source for mititgating the VOC contaminations near ground level or at low altitudes in the atmopshere. As the heights of flare stacks in the examined areas mostly ranging from 100 to 300 m, both saturated and unsaturated hydrocarbons responsible for the VOCs at high altitudes in the atmosphere were expected to be emitted from the local industrial activites. For the influence of a tropical savanna climate, the stronger photochemical reactions and elevating humidity seemed to reduce the VOC concentrations and inhibit the distributions during the daytime and in the wet season. Compared to the influence of diurnal temperature variation on the VOC concentration distribution, the effect of seasonal humidity variation seemed to be more important. In the wet season, aromatic hydrocarbons with more complex structures and high molecular weights and selected aliphatics accounted for the maximum variances of the VOC distributions. For the effect of diurnal temperature variation, the VOCs with stronger photochemical resistances and longer atmoephric lifetimes were the dominant species observed during the daytime, whereas additional VOCs more vulnerable to sunlight or high temperatures also occurred at night.

For the VOCs in a WWTP, the transfers of the aromatics between the water and activated sludge phases were limited, and volatilization from the water to air phases was an important pathway for the their removal in the wastewater. Activated sludge in the bioreactor helped reduce the transfers of aromatic VOCs with higher hydrophobicity from the water to air phases. Xylenes was more readily removed as their concentrations were not found in the anaerobic biological treatment process, whereas benzene were still present at appreciable levels. For the aromatic VOCs with large K_H and small K_{OW} values, aeration more substantially enhanced its volatilization and diminished the biosorption and/or biodegradation in the biological treatment process such as activated sludge. The chlorinated VOCs apeared to be mostly present in the water phase in the treatment processes. The impact of aeration on the performance of biologically treating the chlorinated VOCs was negative, as the aeration actually promoted the VOC emissions into the atmosphere in the treatment process. Regardless of saturated or unsaturated chlorinated VOCs, stronger potentials for biological treatment occurred for those with less chlorine such as chloroform and TCE. While the aeration enhanced the transfers of chlorinated VOCs into the air phase, the VOC removal by volatilization was more important for the two alkane chlorinated VOCs. The alkene chlorinated VOCs were still present and limitedly removed from the water phase.

For the adverse health risks posed by the VOC emissions in a full-scale WWTP, the risks of workers by the exposures to the VOCs in the WWTP were moderate. From the viewpoints of emission rate and health risk, benzene was the species with a higher emission rate and elevated adverse health (both cancer and non-cancer) risks to local workers among the VOCs studied. TCE was the species with a high emission rate but low adverse health (both cancer and non-cancer) risks. The toxicity such as the slope factor and reference concentration of a VOCs was important to determine whether a VOC was of essential concern for the health of local workers. Among different treatment processes, the adverse health risks to workers were relatively higher in the sedimentation process. The VOC emission rates were enhanced by aeration in the aerobic sludge treatment process, but nevertheless, due to the strong mixing and dilution with fresh air by aeration the cancer and non-cancer risks were low. Without aeration causing turbulences at the water surface, the health risks of these VOCs were strongly associated with their emission rates, while the findings were different with aeration as the VOC emission rates were enhanced without significantly elevating the associated adver health risks. In terms of the risks posed to the public living near a WWTP, the areas affected by the occurrences of chlorinated VOCs were relatively larger that those of aromatic VOCs. People living in the areas even more than 4 km away from the WWTP were still potentially exposed to a cancer risk exceeding the regulatory threshold limit, mainly attributed to the presence of chloroform. The discussion here suggested that the VOCs emitted from a WWTP could be one of the important concerns for the chronic adverse health risks for both the workers in the WWTP and those people living some distances away from the plant. These texts offered the key to the complications in the current knowledge regarding the VOC contaminations and hopefully provided insight for managing the adverse impacts of the anthropogenic VOCs on the environment and public health.

REFERENCES

[1] U.S. Geological Survey (USGS). (2016). Volatile Organic Compounds (VOCs) [Webpage]. http://toxics.usgs.gov/definitions/vocs.html.

[2] U.S. Environmental Protection Agency (USEPA). (2016). An introduction to indoor air quality (IAQ) [Webpage]. http://www.epa.gov/iaq/voc.html.

[3] U.S. Environmental Protection Agency (USEPA). (2016). USEPA Integrated Risk Information System (IRIS) Database [Webpage]. http://www.epa.gov/ncea/iris/.

[4] Yang, J. J.; Liu, C. C.; Chen, W. H.; Yuan, C. S.; Lin, C., **2013,** Assessing the altitude effect on distributions of volatile organic compounds from different sources by principal component analysis. *Environmental Science-Processes & Impacts 15* (5), 972-985.

[5] Liu, C. C.; Chen, W. H.; Yuan, C. S.; Lin, C. S., **2014,** Multivariate analysis of effects of diurnal temperature and seasonal humidity variations by tropical savanna climate on the emissions of anthropogenic volatile organic compounds. *Science of the Total Environment 470*, 311-323.

[6] Chen, W. H.; Yang, W. B.; Yuan, C. S.; Yang, J. C.; Zhao, Q. L., **2014,** Fates of chlorinated volatile organic compounds in aerobic biological treatment processes: The effects of aeration and sludge addition. *Chemosphere 103*, 92-98.

[7] Yang, J. C.; Wang, K.; Zhao, Q. L.; Huang, L. K.; Yuan, C. S.; Chen, W. H.; Yang, W. B., **2014,** Underestimated public health risks caused by overestimated VOC removal in wastewater treatment processes. *Environmental Science-Processes & Impacts 16* (2), 271-279.

[8] Chen, W. H.; Yang, W. B.; Yuan, C. S.; Yang, J. C.; Zhao, Q. L., **2013,** Influences of Aeration and Biological Treatment on the Fates of Aromatic VOCs in Wastewater Treatment Processes. *Aerosol and Air Quality Research 13* (1), 225-236.

[9] Yang, W. B.; Chen, W. H.; Yuan, C. S.; Yang, J. C.; Zhao, Q. L., **2012,** Comparative assessments of VOC emission rates and associated health risks from wastewater treatment processes. *Journal of Environmental Monitoring 14* (9), 2464-2474.

[10] U.S. Environmental Protection Agency (USEPA). (2016). Clean Air Act [Webpage]. http://www.epa.gov/air/caa/.

[11] Guo, H.; Lee, S. C.; Chan, L. Y.; Li, W. M., **2004,** Risk assessment of exposure to volatile organic compounds in different indoor environments. *Environmental Research 94* (1), 57-66.

[12] Schwarzenbach, R. P.; Gschwend, P. M.; Imboden, D. M., *Environmental Organic Chemistry*. 2nd Edition ed.; John Wiley & Sons, Inc.: New Jersey, U.S.A., 2003.

[13] Han, Y. M.; Du, P. X.; Cao, J. J.; Posmentier, E. S., **2006,** Multivariate analysis of heavy metal contamination in urban dusts of Xi'an, Central China. *Science of the Total Environment, 355* (1-3), 176-186.

[14] Larsen, R. K.; Baker, J. E., **2003,** Source apportionment of polycyclic aromatic hydrocarbons in the urban atmosphere: A comparison of three methods. *Environmental Science & Technology 37* (9), 1873-1881

[15] Fraley, C.; Raftery, A. E.; Murphy, T. B.; Scrucca, L. *mclust Version 4 for R: Normal Mixture Modeling for Model-Based Clustering, Classification, and Density Estimation*; Department of Statistics, University of Washington: 2012.

[16] U.S. Environmental Protection Agency (USEPA). (1994). Air emissions models for waste and wastewater [Webpage]. http://nepis.epa.gov/Exe/ZyNET.exe/P1005Q16.TXT?ZyActionD=ZyDocument&Client=EPA&Index=1981+Thru+1985&Docs=&Query=&Time=&EndTime=&SearchMethod=1&TocRestrict=n&Toc=&TocEntry=&QField=&QFieldYear=&QFieldMonth=&QFieldDay=&IntQFieldOp=0&ExtQFieldOp=0&XmlQuery=&File=D%3A\zyfiles\Index%20Data\81thru85\Txt\00000015\P1005Q16.txt&User=ANONYMOUS&Password=anonymous&SortMethod=h.

[17] Mattermuller, C.; Gujer, W.; Giger, W., **1981,** Transfer of volatile substances from water to the atmosphere. *Water Research 15* (11), 1271-1279.

[18] Ramaswami, A.; Milford, J. B.; Small, M. J., *Integrated Environmental Modeling: Pollutant Transport, Fate, and Risk in the Environment*. John Wiley & Sons, Inc.: 2005.

[19] California Environmental Protection Agency (CalEPA). (2016). Office of Environmental Health Hazard Assessment (OEHHA) Risk Assessment [Webpage]. http://oehha.ca.gov/risk.html.

[20] Velasco, E.; Lamb, B.; Westberg, H.; Allwine, E.; Sosa, G.; Arriaga-Colina, J. L.; Jobson, B. T.; Alexander, M. L.; Prazeller, P.; Knighton, W. B.; Rogers, T. M.; Grutter, M.; Herndon, S. C.; Kolb, C. E.; Zavala, M.; de Foy, B.; Volkamer, R.; Molina, L. T.; Molina, M. J., **2007,** Distribution, magnitudes, reactivities, ratios and diurnal patterns of volatile organic compounds in the Valley of Mexico during the MCMA 2002 & 2003 field campaigns. *Atmospheric Chemistry and Physics 7*, 329-353.

In: Volatile Organic Compounds
Editor: Julian Patrick Moore
ISBN: 978-1-63485-370-5

Chapter 3

MICROBIAL VOLATILE ORGANIC COMPOUNDS AND TRANS-KINGDOM INTERACTIONS: WHY ARE THEY SO IMPORTANT?

Daniela Minerdi*
Department of Life Sciences and Systems Biology,
University of Torino, Torino, Italy

ABSTRACT

Volatiles are ubiquitously present on Earth due to their physical and chemical properties. Whereas animal and plant volatile emissions have been comprehensively studied in the past, volatiles of microorganisms have been neglected. The wealth of Microbial Volatile Organic Compounds (mVOCs) has been recently discovered. Besides the elucidation of their chemical structures, unraveling the biological functions of mVOCs will be one of the major tasks in the future. Microbial VOCs play important biological roles in multitrophic interactions. The present chapter will illustrate the current knowledge of intra- and inter- organismal mVOC-based interactions, volatile perception, signal transduction and phenotypical responses in the receiver organisms.

---

* Corresponding author: daniela.minerdi@unito.it.

Keywords: microbial volatile organic compounds, symbiosis, quorum sensing, quorum quenching

INTRODUCTION

Communication is the process by which informations are exchanged between individuals through a common system of symbols, signs or behavior. Communicating is something that all animals, bacteria, plants, fungi, insects do. Every organism employs a network of signaling pathways to sense the environment and other organisms and to react with specific molecular, cellular or developmental changes. Sensing of the environment and other organism occurs through the recognition of specific molecules, a vast array of secreted proteins and metabolites play key roles in these mechanisms. Many organisms have evolved the ability to exploit these mechanisms in other organisms to benefit themselves or coordinate symbiosis (Bednarek et al., 2010). The "words" of the chemical conversation of plants and microbes are represented by the Volatile Organic Compounds (VOCs) that affect each other and their environment (Kramer and Abraham, 2012; Morath et al., 2012).

VOCs are low molecular weight (<400 Da) carbon-containing complex mixture of secondary metabolites deriving from different biosynthetic pathways, with a lipophilic character that easily evaporate at normal temperatures and pressures (Bitas et al., 2013). They generally have low to medium water solubility and often have a distinctive odor (Hermann, 2010).

Microbial Volatile Organic Compounds (mVOCs) are a type of VOCs produced by all microorganisms as part of their normal metabolism. These compounds can diffuse a long way from their point of origin; they can persist and migrate in soil environments, area of dense vegetation and other microhabitats. There is a growing recognition of the extent of chemical communication in the biosphere and the role of volatile chemicals play in biological signaling. Chemical signaling takes place within individual organisms, between individuals of the same species and between different species. Due to their ability to move through air space as well as liquids (Effmert et al., 2012), VOCs are ideal signaling molecules for mediating both short and long-distance intercellular and organismal interactions. The ecological role of volatiles in above and belowground interactions among plants, fungi, bacteria and insects is well recognized (Bailly and Weisskopf, 2012; Effmert et al., 2012; Davis et al., 2013; Farag et al., 2013; Audrain et al., 2015; Schmidt et al., 2015; Kanchiswamy et al., 2015a). To date it is known the structural diversity of about 1,700 volatiles from 991 plant species (Dunkel et al., 2009) and 1,093 volatiles from 491 microbes (Lemfack et al., 2014). The term "volatilome" has been proposed to describe their complexity (Maffei et al., 2011).

There are evidences that complex multi-trophic interactions can results from the production of mVOCs. They govern antagonists, pathogenic and mutualistic symbionts (Maffei, 2010; Maffei et al., 2011; Garbeva et al., 2014; Lemfack et al., 2014; Kanchiswamy et al., 2015b).

One of the most fascinating and complex "way of life" diffused in nature is symbiosis. Symbiosis means "living together' and refers to a close and prolonged association between two or more organisms of different species that may last for the lifetime of one or all "partners." Symbioses can be mutualistic (all partners benefiting), commensalistic (one benefiting and the others unharmed), or parasitic. Symbiotic relationships occur between the most distantly related organisms, such as bacteria and eukaryotic cells, as well as between closely related species. Some symbioses are intracellular, some are extracellular, and still others occur between separate partners. Some symbioses are facultative, others obligate, and this characteristic may change through the life cycle of the host. Over evolutionary time, symbiosis has played a vital role in the formation of highly organized life and new types of ecological relationships (Shtark et al., 2010).

Since symbiosis have played and still plays a central role in the evolution, the purpose here is to draw attention to the volatile molecules that regulate the establishment and maintenance of mutualistic and commensalistic symbiosis occurring between plant, fungi, bacteria, algae, insects. Belowground, aboveground and other habitats such as sea will be considered.

MVOC Is One of the Communication Language in Symbiosis

Belowground Communication: The Habitat Rhizosphere and Rhizoplane

Volatiles have attracted attention in belowground communication due to their ability to travel further distances that non-volatiles metabolites (Penuelas et al., 2014). It was only recently recognized that belowground organisms are opulent volatile producers and emitters. It has been extimated that a single gram of soil may contain tens of thousand of different fungal, bacterial, archaeal and protist species. (McNeal and Herbert, 2009). The rhizosphere is an intense habitat in direct proximity of plant roots that includes complex and various root residues and secretions, such as allelochemicals, saccharides, amino acids and growth factors. It is functionally defined as the particulate matter and microorganisms that cling to roots after being gently shaken in water. These substances play important role in root elongation and colonization by rhizospheric microbes, and they are often produced by plant symbionts or affected by plant symbionts. The theoretical extent of the rhizosphere is dependent on the zone of influence of the plant roots and associated microorganisms. The rhizosphere is a metabolically busier, faster moving, and more competitive environment than the surrounding soil. The rhizoplane is the root epidermis and outer cortex where soil particles, bacteria and fungal hyphae adhere (Singer, 2006). The functional definition is the remaining microorganisms and soil particles after the roots have been shaken vigorously in water. There are more microbes in the rhizoplane than in the more loosely associated rhizosphere. Bacteria and fungi that live within the cells of the root are not considered a part of the rhizoplane, but instead called endophytes (Sylvia, 2005). Communication with rhizosphere bacterial populations, mycelial colonies, and between fungi and bacteria is mediated by signaling molecules, but we are only beginning to learn this specific language. In this chapter will be presented an overview of the deciphered codes that govern bi-trophic and tripartite trans-kingdom interactions.

BACTERIA-PLANT SYMBIOSIS

Root Nitrogen Fixing Nodules

Plants are colonized by many types of symbionts. One of the best characterized symbiosis is the one taking place between nitrogen fixing bacteria and more than 15,000 plant species from at least 12 families (Sprent, 2001); with the help of symbiotic nitrogen fixing rhizobia in their root nodules, most legumes can grow in poor soils without the addition of nitrogen-containing fertilizers.

Medicago truncatula is a model plant for understanding legume–bacteria interactions. *M.truncatula* roots form a specific root–nodule symbiosis with the nitrogen-fixing bacterium *Sinorhizobium meliloti*. Symbiotic nitrogen fixation generates high Fe demands for bacterial nitrogenase holoenzyme and plant leghemoglobin proteins. Leguminous plants acquire Fe via rhizosphere acidification and enhanced ferric reductase activity. Orozco-Mosqueda and colleagues (2013) tested the ability of *S. meliloti* to promote *M. truncatula* plant growth via the emission of mVOCs. They evaluated the capability of the plant and bacteria to interact separately via VOCs without physical contact. The profile of *S. meliloti* VOCs consisted of a mixture of six compounds found in axenic cultures, all of them were also present in the plant–bacterium interaction system. Analogously, the plants produced a mixture of seven VOCs that were present in axenic cultures, six of which were also being present in the interacting system. Since these latter compounds did not all occur in bacterial cultures, a plant origin was indicated. Furthermore, a mixture of five VOCs was also detected only in the plant– bacterial interaction systems. This shows that the plant, bacterium, or both organisms, may detect its symbiotic partner, and modify its metabolism to produce new volatile compounds. The mVOC mixture detected from the *M. truncatula-S. meliloti* interaction contains hexadecylamine that is able to induce rhizosphere acidification and Fe-reduction activity in *M. truncatula* (Orozco-Mosqueda et al., 2013). The rhizosphere acidification capacity of plants induced with bacterial VOCs, under both Fe-rich and-deficient conditions, was investigated in comparison with non-induced controls. Plants induced with bacterial VOCs exhibited a better ability to acidify their rhizospheric environment compared to non-induced plants even when Fe was abundant. Non-induced plants placed in Fe-rich medium produced very limited acidification of the rhizosphere. The mixture of VOCs produced by *S. meliloti* also promoted a significant increase in *M. truncatula* chlorophyll concentrations and in plant biomass. This improved status may indicate that the plant is preparing to support the establishment of the nodule symbiosis with *S. meliloti.*

Recent data pointed out that associations between plants and mutualistic soil microbiota not only influence plant growth, but also affect the outcome of interactions between plants and organisms of higher trophic levels such as herbivores and pathogens (Bonte et al., 2010; Pineda et al., 2010).

The provision of additional nitrogen to plants by rhizobia is expected to affect plant interactions with herbivores. Plants generally produce a wide range of VOCs, which comprise mainly fatty acid derivatives, terpenoids, phenyl propanoids, and benzenoids (Winter and Rosta´s 2010; Ballhorn et al. 2011). Many of these volatiles are induced by herbivores being synthesized *de novo* in response to feeding damage triggering multiple functions in plant–animal and plant– plant interaction (Baldwin, 2010). One of the best-studied effects of VOCs is their function as an indirect plant defense; they provide olfactory cues that attract parasitoid wasps to the damaged VOC-emitting plant (Rosta´s and Turlings, 2008). The parasitoids lay their eggs into herbivore bodies the wasps' larvae feed on inner tissues, kill the herbivore reducing its pressure on the plant (Digilio et al., 2010).

Therefore, any influences of belowground symbionts on volatile production may translate into changes of aboveground interactions. The importance of rhizobia in plant–herbivore interactions is still in its infancy (Ballhorn et al., 2009). Rhizobia are able to change the composition of the jasmonic acid-induced volatile blend of lima bean (Ballhorn et al., 2013). Within these VOC blends, individual compounds were differently affected. While the concentrations of indole was significantly increased in induced rhizobia-colonized lima bean plants the majority of C-based volatiles remained unaffected (2-ethylhexan-1-ol,cis-b-ocimene, linalool, cis-jasmone) or were decreased by 59% (cis-3-hexenyl acetate), 56% (cis-3- hexenyl butyrate and cis-3-hexenyl isovalerate), and 38% (β-caryophyllene]. The total volatile emission after induction was reduced for rhizobial plants and the volatile blend was dominated by indole. The effects of VOCs produced by rhizobia colonized and not colonized plants on the specialist insect herbivore Mexican bean beetle was tested in olfactometer experiments with artificially enhanced indole levels in otherwise natural VOC blends. Rhizobia colonization has a substantial impact on the emission of inducible plant volatiles and host plant selection by an insect herbivore. Results suggest a key function of this compound in herbivore deterrence. This study demonstrated a rhizobia-triggered efficacy of induced plant defense via volatiles. suggesting rhizobia as an integral part of legume defenses against herbivores (Ballhorn et al., 2013).

Rhizobia produce oligosaccharides called nodulation factors (Nod factors) to communicate their presence to the host plant and induce nodulation response in the plant. Leitner and colleagues (2010) proved that *Medicago truncatula* reacts to these elicitors with the emission of volatiles sesquiterpenes, a typical defence-related reaction.

ENDOPHYTES

Endophytes are symbiotic bacteria, fungi or actinomycetes that live intercellularly within the tissue of almost all plants (Bacon and White, 2000). They exhibit complex interactions with their hosts, which involve mutualism and antagonism. Plants strictly limit the growth of endophytes, and these endophytes use many mechanisms to gradually adapt to their living environments (Dudeja et al., 2012). In order to maintain stable symbiosis, endophytes produce several compounds that promote growth of plants and help them to adapt better to the environment (Das and Varma, 2009).

Endophytes produce a wide range of compounds useful for plants for their growth, protection to environmental conditions, and sustainability, in favor of a good dwelling place within the hosts. They protect plants from herbivores by producing compounds which prevent animals from further grazing on the same plant and sometimes act as biocontrol agents.

Enterobater aerogenes is an endophytic bacterium able to colonize the aboveground parts of healthy maize seedlings producing the volatile 2,3-butanediol. This mVOC increases pathogen resistance of the host plant, increases the growth of the larvae of the herbivorous *Spodoptera littoralis* and does not promote plant growth. The altered state of the plant is shown to be able to affect the interaction with the parasitic wasp *Cotesia marginiventris*, a predator of the *S. littoralis* larvae (D'Alessandro et al., 2014).

Fungal endophytes have been studied as a source of novel secondary metabolites (Tan and Zou 2001) and also have gained attention as producers of bioactive mVOCs. *Muscodor albus* is an endophyte that produces a blend of VOCS that are inhibitory or lethal to a wide range of bacteria and pathogenic fungi. *M. albus* produced a mixture of volatile acids, alcohols, esters, ketones, and lipids, which individually had inhibitory but not lethal effects against test species such as *Fusarium solani, Pythium ultimum*, and *Rhizoctonia solani* (Strobel et al., 2001). When applied collectively, these same VOCs acted synergistically to kill a broad range of plant pathogenic fungi and bacteria.

Since the original isolation of *Muscodor albus* from a cinnamon tree, several other *Muscodor* strains and species that emit antibiotic mixtures of VOCs have been isolated (Zhang and Li, 2010). This selective antimicrobial effect can be harnessed against undesirable pathogens and has been termed "mycofumigation."

Fungi-Plant Symbiosis

Arbuscular mycorrhizas (AMs) are some of the most widespread symbioses on earth. More than 80% of plants are known to establish AM associations with fungi from the Glomeromycota genus, which improves the mineral nutrition of plants, particularly of phosphorus (Smith and Read, 2008). In return, plants provide sugar to the fungal symbionts. The establishment of an AM symbiosis proceeds as a series of genetically controlled steps and commences with a pre-symbiotic molecular crosstalk leading to reciprocal perception (Nadal and Paszkowski, 2013). The fungus makes contact with the root surface via a hyphopodium only if successful recognition occurs. The hyphopodium then penetrates into the root leading to the formation of arbuscules, or sites of nutrient exchange (Nadal and Paszkowski, 2013). It is known that the AM fungal spores germinate spontaneously. During this period, the hyphae respond to several root-secreted chemicals, resulting in distinct growth patterns. It has been suggested that AM fungi have perception machinery enabling the differential recognition of chemically diverse plant metabolites and the transduction of the information into compound-specific morphological responses (Nadal and Paszkowski, 2013). These early responses are likely to play crucial roles in the preparation for a successful colonization process. Plants can also sense and respond to signals released by AM fungi. Lateral root formation has been demonstrated as a common response to the presence of germinating spore exudates and germinated spores emitted VOCs in some plant species (Olah et al., 2005; Mukherjee and Ane, 2011). The study of the expression profile of the genes involved in mycorrhization establishment and root development in *Lotus japonicus* showed that only *LfCCD*7 gene that code for an important component of the strigolactone synthetic pathway, was differentially expressed following exposure to the volatiles produced by germinated spores (Sun et al., 2015).

There is also evidence that roots emit volatile signals that stimulate the directional growth of the AM fungus toward them (Gemma and Koske, 1988). Furthermore, it has been suggested that AMs could alter the profile of the volatile organic compounds released by a plant, although the mechanisms have still to be elucidated (Schausberger et al., 2012; Babikova et al., 2013).

AM fungi not only have underground interactions with the roots of host plants but, in an indirect way, they also exhibit above-ground interactions with aphids. Aphids are among the most abundant and agriculturally important invertebrate herbivores (Minks and Herrewijn, 1989). They feed on plant sap directly from the phloem, draining the plant of nutrient resources and greatly reducing plant fitness and biomass (Guerrieri and Digilio 2008). Therefore, there is considerable potential for interactions between AM fungi and aphids competing for plant nutrients. Insect herbivores induce systemic defense-related signaling in host plants such as the salicylic acid and jasmonic acid signalling pathways (Goggin, 2007), which affect the biosynthesis of plant VOCs. Therefore, the type and quantity of VOCs can change significantly when plants are attacked by herbivores (Unsiker et al., 2009; Dicke 2009), becoming less attractive or repellent to subsequent herbivores (Dicke 1999), and attractive to natural enemies of these herbivores, such as parasitoids (Turlings et al., 1995). Salicylic acid and jasmonic acid signaling pathways are also regulated by mycorrhizal colonization in order for AM fungi to achieve compatibility with host plants (Pozo and Azcon- Aguilar, 2007). Therefore, AM fungi may also affect the biosynthesis of VOCs and consequently aphid host location.

Guerrieri and co-workers (2004) found that mycorrhizal tomato were more attractive to parasitoids enemies of aphids than were non-mycorrhizal plants. Schausberger et al., (2012) found that AM fungi affected the VOCs emitted by bean plants making them more attractive to predators of spider mites. A recent study showed that AM fungi increased the attractiveness of plants to aphids. AM fungi suppressed emission of the sesquiterpenes (E)-caryophyllene and (E)-b-farnesene, and aphid attractiveness to VOCs is associated with the proportion of sesquiterpenes in the sample. Emission of (Z)-3-hexenyl acetate, naphthalene and (R)-germacrene D was regulated by an interaction between aphids and AM fungi. Aphids had a negative effect on mycorrhizal colonization, plant biomass and nutrition. AM fungi have a key bottom-up role in insect host location by increasing the attractiveness of plant VOCs to aphids, whereas aphids inhibit formation of AM symbioses.

Ectomycorrhiza

Trufles (*Tuber* spp.) are fungi that forms mutualistic association called ectomycorrhiza with the roots of plant, that promote plant health. These fungi use VOCs throughout their life to regulate their interactions with other organisms. More than 200 volatiles have been described from various truffle species in the presymbiotic mycelial stage, during the mycorrhizal stage when the fungus enters in a symbiosis with plant roots and during the reproductive stage (Splivallo et al., 2011).

A few of these volatiles affect the root architecture of plants under laboratory conditions, resulting in primary root shortening (Splivallo et al., 2007) and root hair elongation (Splivallo et al., 2009).

Ditengu and colleagues (2015) identified sesquiterpenes produced by *Laccaria bicolor* as biologically active agents while interacting with *Populus* and *Arabidopsis.* Inhibition of fungal sesquiterpenes production by lovastatin strongly reduces lateral roots proliferation. The lateral roots promoting sesquiterpenes signal enhance the root surface area for plant nutrient uptake and improve fungal access to plant-derived carbon via root exudates.

FUNGI-BACTERIA SYMBIOSIS: TRIPARTITE INTERACTIONS

To date, vey little is known about the role of fungal and bacterial volatiles in fungal-bacterial interactions. In the case of an ecto/endo symbiotic association a new biological entity is formed deriving by the "fusion" of the prokaryotic and eukaryotic partner. This new entity produces its own type of VOCs that are different from the ones produced by the two single partners, these molecules can also have an effect on other soil organisms including bacteria, fungi and plants in a tripartite very complex communication. One example of this situation is represented by the antagonistic fungus *Fusarium oxysporum* MSA35 isolated from an Italian suppressive soil. *Fusarium oxysporum* MSA35 [wild-type (WT) strain] is an antagonistic *Fusarium* that lives in association with a consortium of bacteria belonging to the genera *Serratia, Achromobacter, Bacillus* and *Stenotrophomonas.* Typing experiments and virulence tests provided evidence that the *F. oxysporum* isolate when cured of the bacterial symbionts [the cured (CU) form], is pathogenic, causing wilt symptoms identical to those caused by *F. oxysporum* f. sp. *lactucae* (Minerdi et al., 2008). VOCs emitted from the WT strain negatively influence the mycelial growth of different pathogenic *formae speciales* of *F. oxysporum.* Furthermore, these VOCs repress gene expression of two putative virulence genes in *F. oxysporum lactucae* strain Fuslat10, a fungus against which the WT strain MSA 35 has antagonistic activity. The VOC profile of the WT and CU fungus shows different compositions. Sesquiterpenes, mainly caryophyllene, were present in the headspace only of WT MSA 35. No sesquiterpenes were found in the volatiles of ectosymbiotic *Serratia* sp. strain DM1 and *Achromobacte*r sp. strain MM1. Bacterial volatiles had no effects on the growth of the different ff. spp. of *F. oxysporum* examined. Hyphae grown with VOC from WT *F. oxysporum* f. sp. *lactucae* strain MSA 35 were hydrophobic whereas those grown without VOCs were not, suggesting a correlation between the presence of volatiles in the atmosphere and the phenotype of the mycelium. Authors proposed a new potential direct long distance mechanism for antagonism by reduction of pathogen mycelial growth and inhibition of pathogen virulence gene expression (Minerdi et al., 2009). Further research found that the WT strain promoted lettuce growth and expansin A5 gene expression through microbial VOC emissions. β-caryophyllene was found to be the main component of the volatiles released by the WT strain responsible for the plant growth promotion effect (Minerdi et al., 2011).

Another study of tripartite interactions is the one of Son and co-workers (2009), they showed that *Paenibacillus polymyxa* and *Paenibacillus lentimorbus* exhibited strong antifungal activities, thereby interfering with the interactions between *Meloidogyne incognita* and *Fusarium oxysporum* and concomitant nematode infestation of tomato plants.

An endobacterium of the phylum *Cytophaga-Flexibacter-Bacteroides* has been reported in pure mycelial cultures of *Tuber borchii* (Barbieri et al., 2000). Whether these microbes contribute to the characteristic aroma of truffles is open to speculation. Indirect proof may nevertheless support this hypothesis. Some volatile such as 2-methyl-4,5-dihydrothiophene, specific to *T. borchii* fruiting bodies, have not been found in pure mycelial cultures of the same species (Splivallo et al., 2007). These findings suggest that either the volatiles of the fruiting bodies are under strict developmental control or they are produced by associated microbes or even by both microbes and truffles, one using a precursor produced by the other. In conclusion, the role of microbes associated with truffle fruiting bodies as producers of the components of truffle is under investigation.

MVOCs and Insect Symbiosis

Aphids are one of the most serious pests of crops worldwide, causing major yield and economic losses. When feeding on the plant phloem sap, aphid excretes large amounts of honeydew, a complex mixture of sugars, organic acids, amino acids and some lipids (Leroy et al., 2011a). The composition of honeydew is driven by the host plant that the aphid is feeding on but also by the aphid itself, including its primary symbionts. Essential amino acids are also provided by endosymbiotic bacteria to balance the concentrations of amino acids in the phloem sap (Febvay et al, 1999). This excretory product is considered as a food complement for many aphid natural enemies (Hogervorst, 2007) and the volatiles that are released are supposed to act as kairomones for parasitoids and predators (Choi et al., 2004; Verheggen, 2008; Almohamad, 2009). Because honeydew is mainly composed of sugars and amino acids, this excretion constitutes an excellent growth medium for microorganisms that actively contribute to the production of volatile compounds· Leroy and colleagues (2011) isolated *Staphylococcus sciuri,* an endosymbiotic bacterium of the pea aphid *Acyrthosiphon pisum* honeydew, that produces semiochemicals volatile compounds that act as effective attractants and ovipositional stimulants for the aphidophagous predator *Episyrphus balteatus*.

Other Type of Symbiosis: Lichens and Corals

Lichens are a symbiotic association of a fungus and a chlorophyll containing partner, either green algae or cyanobacteria, or both. The fungus provides a suitable habitat for the partner, which provides photosynthetically fixed carbon as energy source for the system. The evolutionary result of the self-sustaining partnership is a unique joint structure, the lichen thallus, which is indispensable for fungal sexual reproduction. The classical view of a dual symbiosis has been challenged by recent microbiome research, which revealed host-specific bacterial microbiomes. Multi-omics approaches have provided evidence for functional contribution by the bacterial microbiome to the entire lichen meta-organism while various abiotic and biotic factors can additionally influence the bacterial community structure (Aschenbrenner et al., 2016). It has been proved that boreal lichens can exchange oxygenated volatile organic compounds with the atmosphere (Kesselmeier et al., 1999).

Corals are symbiosis between an animal, the polyp and green algae, the zooxantelle. Soft corals belonging to the genus *Cespitularia* have been well recognized as a rich source of bioactive secondary metabolites especially diterpenoids, sesquiterpenoids (Elshamy et al., 2016). Both lichens and corals have been recognized to produce VOCs but their function in the biology of the symbiotic relationship is still unknown.

Two Communication Systems Overlapping: Volatile Organic Compounds and Quorum Sensing Molecules

Quorum sensing is a cell-cell communication mechanism by which bacteria count their own numbers by producing and detecting the accumulation of a signaling molecule that they export into their environment. As a population of quorum-sensing bacteria grows, a proportional increase in the extracellular concentration of the signaling molecule occurs. When a threshold concentration is reached, the group detects the signaling molecule and responds to it with a population-wide alteration in gene expression. Processes controlled by quorum sensing are usually ones that are unproductive when undertaken by an individual bacterium but become effective when undertaken by the group (Bassler and Losick, 2006). The first quorum-sensing system described at the molecular level was in *Vibrio fischeri,* a bioluminescent marine bacterium in which the bacteria grow to a high cell density within the light organ of the Hawaiian squid (Nealson and Hastings, 1979). When *V. fischeri* cells are free living, they do not luminesce. However, in the nutrient rich environment of the light organ of the squid, they grow to high concentration and produce an acyl-homoserine lactone which acts as autoinducer (Eberhard et al., 1981). The population-dependent release of autoinducer leads to transcription of the luciferase gene and bioluminescence (Kaplan and Greenberg, 1985). Gram-negative bacteria cell to cell signaling systems use fatty acid derivatives as autoinducers (Bassler and Losick, 2006). Quorum sensing also occurs widely in Gram-positive bacteria, where the best studied autoinduction signals are oligopeptides (Waters and Bassler, 2005).

Quorum sensing molecules regulate virulence and biofilm formation in *Pseudomonas aeruginosa* (Rumbaugh et al., 2000), fruiting body production in *Myxococcus* (Shimkets, 1999), antibiotic and pigment production in *Serratia* (Thomson et al., 2000), pathogenicity in plant-pathogenic bacteria (von Bodman et al, 2003).

There is increase evidence that quorum sensing is widespread also in fungi. The best studied system is *Candida albicans* a dimorphic fungus, which grows as a commensal on humans in yeast form but functions as an opportunistic pathogen. The yeast–hypha transition is essential for causing disseminated disease. Hyphal development is suppressed by farnesoic acid, which acts as a quorum-sensing molecule (Hogan, 2006).

Quorum sensing is usually discussed and studied in the context of aqueous environments. Some environmental conditions, for example those found in soil, may not always be ideal for signaling in the liquid phase. Local cell numbers may be quite high, but not linked to one another through the liquid phase. The production of signals able to act through the gas phase would circumvent these issues. Only a few example of volatile quorum sensing signals have been discovered: *Ralstonia solenacearum,* a soil-borne plant pathogen, uses the volatile signal 3-hydroxyl palmitic acid methyl ester, for regulating the expression of most of the traits needed for infection and virulence (von Bodman et al., 2003). The search of soil mVOCs acting as autoinducers is at the beginning and it is highly likely that many other volatile molecules will be discovered and that signaling in the gas phase may be one of the next important frontiers in quorum sensing (Horswill et al., 2007).

On the other hand VOCs produced by rhizospheric bacteria have been shown to inhibit quorum sensing. The ability to disrupt quorum sensing networks, termed quorum quenching (QQ), is an important mechanism of competition between bacteria, which may give one bacterial species an advantage over the other. QQ can be achieved by several approaches, including inhibition or inactivation of AHL production (Boyer and Wisniewski-Dye, 2009; Uroz et al., 2009). AHL-mediated communication between individual bacterial cells has been detected in the rhizosphere (Steidle et al., 2001), and rhizospheric bacteria have been shown able to persist and produce VOCs inside the plant (Dandurishvili et al., 2011). Chernin and colleagues (2011) showed that VOCs produced by rhizospheric strains *Pseudomonas fluorescens* and *Serratia plymuthica* can inhibit *N*-acylhomoserine lactone mediated quorum sensing produced by *Agrobacterium*, *Chromobacterium*, *Pectobacterium* and *Pseudomonas.* This quorum-quenching effect was observed when AHL-producing bacteria were treated with VOCs emitted by *P. fluorescens* and *S. plymuthica.* This treatment caused a decrease in the amount of AHLs produced by the bacteria and a suppression of transcription of AHL synthase genes (Chernin et al., 2011).

CONCLUSION

John Donne wrote: “No man is an island.” On earth might exist 10^9 bacterial species (Schloss and Handelsam, 2004), 2,98,000 plant species (Mora et al., 2011) and 1,5 million fungal species (Hawksworth, 2001). The number of volatiles will increase as new species will be discovered and characterized; it appears that most of the organisms are not either an island but instead weave intimate and complex relationships with other forms of life. The amazing with mVOCs is that they can function both as gaseous quorum sensing messengers in soil and as quorum sensing inhibiting molecules. Furthermore, microbial volatile compounds have a preeminent role in the establishment of bacteria-plant, bacteria-fungi symbiosis and at the same time, they directly influence the host defense system and, in indirect way, the interaction with herbivores predators. Taken together all these properties will allow in the near future the exploitation of mVOCs as an eco-friendly, cost-effective, and sustainable strategy for agricultural practices.

REFERENCES

Almohamad, R., Verheggen, F. J. And Haubruge, E. (2009). Searching and oviposition behavior of aphidophagous hoverflies (Diptera: Syrphidae): a review. *Biotechnol. Agron. Soc. Environ.*13: 467-481.

Aschenbrenner, I.A., Cernava, T., Berg, G. and Grube, M. (2016). Understanding Microbial Multi-Species Symbioses. *Front. Microbiol.* 18:180.

Audrain. B., Farag, M.A., Ryu, C.M. and Ghigo, M. (2015). Role of bacterial volatile compounds in bacterial biology. *FEMS Microbiol. Rev.* 39: 222-33.

Babikova, Z., Gilbert, L., Bruce, T., Dewhirst, S.Y., Pickett, J.A. and Johnson, D. (2013). Arbuscular mycorrhizal fungi and aphids interact by changing host plant quality and volatile emission. *Funct. Ecol.* 28: 375-385.

Bacon, C.W. and White, J.W. (2000). Microbial endophytes. Dekker, New York.

Bailly, A. and Weisskopf, L. (2012). The modulating effect of bacterial volatiles on plant growth: current knowledge and future challenges. *Plant Signal. Behav.* 7:79-85.

Baldwin, I.T. (2010). Plant volatiles. *Curr Biol* 20:392–397.

Ballhorn, D.J., Kautz, S. and Schädler, M. (2013). Induced plant defense via volatile production is dependent on rhizobial symbiosis. *Oecologia.* 172:833-46.

Ballhorn, D.J., Reisdorff, C. and Pfanz, H. (2011). Quantitative effects of enhanced $CO2$ on jasmonic acid induced plant volatiles of lima bean (*Phaseolus lunatus L.*). *J. Appl. Bot. Food Qual.* 84: 65–71.

Ballhorn, D.J., Kautz, S., Hei, M. and Hegeman, A.D. (2009). Analyzing plant defenses in nature. *Plant Signal. Behav.* 4:743–745.

Barbieri. E., Potenza, L., Rossi, I., Sisti, D., Giomaro, G., Rossetti, S., Beimfohr, C. and Stocchi, V. (2000). Phylogenetic characterization and in situ detection of a *Cytophaga–Flexibacter-Bacteroides* phylogroup bacterium in *Tuber borchii* Vittad. ectomycorrhizal mycelium. *Appl. Environ. Microbiol.* 66: 5035–5042.

Bassler, B.L. and Losick, R. (2006). Bacterially speaking. *Cell* 125:237–246.

Bednarek, P., Kwon, C., and Schulze-Lefert, P. (2010). Not a peripheral issue: secretion in plant–microbe interactions. *Curr. Opin. Plant Biol.* 13: 378-387.

Bitas,V., Kim, H.S., Bennett, J.W., and Kang, S. (2013). Sniffing on microbes: diverse roles of microbial volatile organic compounds in plant health. *Mol. Plant Microbe Interact.* 26: 835-843.

Bonte, D., de Roissart, A., Vandegehuchte, M.L., Ballhorn, D.J. and de la Pena, E. (2010). Local adaptation of aboveground herbivores towards plant phenotypes induced by soil biota. *PLoS ONE* 5:e11174.

Boyer, M. and Wisniewski-Dye, F. (2009). Cell-cell signalling in bacteria: not simply a matter of quorum. *FEMS Microbiol. Ecol.* 70: 1–19.

Chernin, L., Toklikishvili, N., Ovadis, M., Kim, S., Ben-Ari, J., Khmel, I. and Vainstein, A. (2011). Quorum-sensing quenching by rhizobacterial volatiles. *Env. Microbiol. Reports* 6: 698–704.

Choi, M.Y., Roitberg, B.D., Shani, A., Raworth, D. A. and Lee, G.H. (2004). Olfactory response by the aphidophagous gall midge *Aphidoletes aphidimyza* to honeydew from green peach aphid, *Myzus persicae. Entomol. Exp. Appl.* 111: 37–45.

D'Alessandro, M., Erb, M., Ton, J., Brandenburg, A, Karlen, D., Zopfi, J. and Turlings, T.C. (2014). Volatiles produced by soil-borne endophytic bacteria increase plant pathogen resistance and affect tritrophic interactions. *Plant Cell Environ.* 37:813-26.

Dandurishvili, N., Toklikishvili, N., Ovadis, M., Eliashvili, P., Giorgobiani, N., Keshelava, R., et al. (2011). Broad-range antagonistic rhizobacteria *Pseudomonas fluorescens* and *Serratia plymuthica* suppress *Agrobacterium* crown-gall tumors on tomato plants. *J. Appl. Microbiol.* 110: 341–352.

Das, A. and Varma, A. (2009). Symbiosis: the art of living. In Symbiotic Fungi Principles and Practice, A. Varma and A. C. Kharkwal, Eds., pp. 1–28, Springer, Berlin, Germany.

Davis, T.S., Crippen, T.L., Hofstetter, R.W. and Tomberlin, J.K. (2013). Microbial volatile emissions as insect semiochemicals. *J. Chem. Ecol.* 39: 840–859.

Dicke, M. (2009). Behavioural and community ecology of plants that cry for help. *Plant Cell Environ.* 32: 654–665.

Digilio, M.C., Corrado, G., Sasso, R., Coppola, V., Iodice, L., Pasquariello, M., Bossi, S., Maffei, M.E., Coppola, M., Pennacchio, F., Rao, R. and Guerrieri, E. (2010). Molecular and chemical mechanisms involved in aphid resistance in cultivated tomato. *New Phytol.* 87: 1089–1101.

Ditengou. F.A., Müller, A., Rosenkranz, M., Felten J., Lasok, H., van Doorn, M.M., Legué, V., Palme, K., Schnitzler, J.P. and Polle, A. (2015). Volatile signalling by sesquiterpenes from ectomycorrhizal fungi reprogrammes root architecture. *Nat. Commun.* 23: 6279.

Dudeja, S.S., Giri, R., Saini, R., Suneja-Madan, P. and Kothe, W. (2012). Interaction of endophytic microbes with legumes. *J. Basic Microbiol.* 52: 248–260.

Dunkel, M., Schmidt, U., Struck, S., Berger, L., Gruening, B., Hossbach, J. et al. (2009). SuperScent-a data base of flavors and scents. *Nucleic Acids Res.* 37: D291-D294.

Eberhard, A., Burlingame, A.L., Eberhard, C., Kenyon, G.L., Nealson, K.H. and Oppenheimer, N.J. (1981). Structural identification of autoinducer of *Photobacterium fischeri* luciferase. *Biochemistry* 20: 2444–2449.

Effmert, U., Kalderás, J., Warnke, R., and Piechulla, B. (2012). Volatile mediated interactions between bacteria and fungi in the soil. *J. Chem. Ecol.* 38: 665–703.

Elshamy, A.I., Nassar, M.I., Mohamed, T.A. and Hegazy, M.E. (2016). Chemical and biological profile of *Cespitularia* species: A mini review. *J. Adv. Res.* 7: 209-24.

Farag, M.A., Zhang, H., and Ryu, C.M. (2013). Dynamic chemical communication between plants and bacteria through airborne signals: induced resistance by bacterial volatiles. *J. Chem. Ecol.* 39: 1007–1018.

Febvay, G., Rahbe, Y., Rynkiewicz, M., Guillaud, J. and Bonnot, G. (1999). Fate of dietary sucrose and neosynthesis of amino acids in the pea aphid, Acyrthosiphon pisum, reared on different diets. *J. Exp. Biol.* 202: 2639–2652.

Garbeva, P., Hordijk, C., Gerards, S. and de Boer, W. (2014). Volatiles produced by the mycophagous soil bacterium *Collimonas. FEMS Microbiol. Ecol.* 87 639–649.

Gemma, J.N. and Koske, R.E. (1988). Pre-infection interactions between roots and the mycorrhizal fungus *Gigaspora gigantea:* chemotropism of germ-tubes and root growth response. *Trans. Br. Mycol. Soc.* 91:123-132.

Goggin, F.L. (2007). Plant–aphid interactions: molecular and ecological perspectives. *Curr. Opin. Plant. Biol.* 10: 399–408.

Guerrieri, E., Lingua, G., Digilio, M.C., Massa, N. and Berta, G. (2004). Do interactions between plant roots and the rhizosphere affect parasitoid behaviour? *Ecol. Entomol.* 29:753–756.

Guerrieri, E. and Digilio, M.C. (2008). Aphid–plant interactions: a review. *J. Plant Interact.* 3: 223–232.

Hawksworth, D.L. (2001). The magnitude of fungal diversity: the1·5 million species estimate revisited. *Mycol. Res.* 105: 1422–1432.

Hermann, A. (2010). The chemistry and biology of volatiles. *Wiley*, Chichester.

Hogan, D.A. (2006). Talking to themselves: autoregulation and quorum sensing in fungi. *Eukaryot Cell* 5:613–619.

Hogervorst, P., Wäckers, F.L. and Romeis, J. (2007). Effects of honeydew sugar composition on the longevity of *Aphidius ervi. Entomol. Exp. Appl.* 122: 223–232.

Horswill, A., Stoodley, P., Stewart, P. and Parsek, M. (2007). The effect of the chemical, biological, and physical environment on quorum sensing in structured microbial communities. *Anal. Bioanal. Chem*. 387:371-380.

Kanchiswamy, C.N., Malnoy M. and Maffei, M.E. (2015a).Chemical diversity of microbial volatiles and their potential for plant growth and productivity. *Front Plant Sci.* 13;6:151.

Kanchiswamy, C.N., Malnoy, M. and Maffei. M.E. (2015b). Bioprospecting bacterial and fungal volatiles for sustainable agriculture. *Trends Plant Sci.* 20: 206-11.

Kaplan, H.B. and Greenberg, E.P. (1985). Diffusion of autoinducer is involved in regulation of the *Vibrio fischeri* luminescence system. *J. Bacteriol.* 163:1210–1214.

Kesselmeier, J., Wilske, B., Muth,S., Bode, K. and Wolf, A. Exchange of Oxygenated Volatile Organic Compounds Between Boreal Lichens and the Atmosphere. In: Laurila, T. and Lindfors,V. 1999). European Commission: Biogenic VOC emissions and photochemistry in the boreal regions of Europe –Biphorep.

Kramer, R., and Abraham, W.R. (2012). Volatile sesquiterpenes from fungi: What are they good for? *Phytochem. Rev*. 11:15-37.

Leitner, M., Kaiser, R., Rasmussen, M.O, Driguez, H., Boland, W. and Mithofer, A. (2008). Microbial oligosaccharides differentially induce volatiles and signalling components in *Medicago truncatula. Phytochemistry*. 69: 2029-40.

Lemfack, M. C., Nickel, J., Dunkel, M., Preissner, R. and Piechulla, B. (2014). mVOC: a database of microbial volatiles. *Nucleic Acids Res*. 42 D744–D748.

Leroy, P.D., Wathelet, B., Sabri, A., Francis, F., Verheggen, F.J., Capella, Q., Thonart, P. and Haubruge, E. (2011a). Aphid-host plant interactions: does aphid honeydew exactly reflect the host plant amino acid composition? *Arthropod-Plant Inte*. 5: 1-7.

Leroy, P.D., Sabri, A., Heuskin, S., Thonart, P., Lognay, G., Verheggen, F.J., Francis, F., Brostaux, Y., Felton, G.W. and Haubruge, E. (2011b).

Microorganisms from aphid honeydew attract and enhance the efficacy of natural enemies. *Nat. Commun.* 348: 2-7.

Maffei, M. (2010). Sites of synthesis, biochemistry and functional role of plant volatiles. *S. Afr. J. Bot.* 76: 612–63.

Maffei, M. E., Gertsch, J. and Appendino, G. (2011). Plant volatiles: production, function and pharmacology. *Nat. Prod. Rep.* 28: 1359-1380.

McNeal, K.S. and Herbert, B.E. (2009). Volatile organic metabolites as indicators of soil microbial activity and community composition shifts. *Soil. Sci. Soc. Am. J.* 73:579–588.

Minerdi, D,. Moretti, M., Gilardi, G., Barberio, C., Gullino, M.L. and Garibaldi, A. (2008). Bacterial ectosymbionts and virulence silencing in a *Fusarium oxysporum* strain. *Environ. Microbiol.* 10: 1725–1741.

Minerdi, D., Bossi, S., Gullino, M.L. and Garibaldi, A. (2009). Volatile organic compounds: a potential direct long-distance mechanism for antagonistic action of *Fusarium oxysporum* strain MSA 35. *Environ. Microbiol.* 11:844–854.

Minerdi, D., Bossi, S., Maffei, M.E., Gullino, M.L. and Garibaldi, A. (2011). Fusarium oxysporum and its bacterial consortium promote lettuce growth and expansin A5 gene expression through microbial volatile organic compound (MVOC) emission. *FEMS Microbiol. Ecol.* 76:342-351.

Minks, A.K. and Herrewijn, P. (1989). Aphids: Their Biology, Natural Enemiesand Control. Elsevier, New York, New York, USA, pp. 450.

Mora, C., Tittensor, D.P., Adl, S., Simpson, A.G.B. and Worm, B. (2011). How many species are there on Earth and in the Ocean? *PLoSBiol.* 9:e1001127.

Morath, S., Hung, R., and Bennett, J.W. (2012). Fungal volatile organic compounds: A review with emphasis on their biotechnological potential. *Fungal Biol. Rev.* 30:1-11.

Mukherjee, A. and Ane, J.M. (2011). Germinating spore exudates from arbuscular mycorrhizal fungi: molecular and developmental responses in plants and their regulation by ethylene. *Mol. Plant Microbe Interact.* 24: 260-270.

Nadal, M., Paszkowski, U. (2013). Polyphony in the rhizosphere: presymbiotic communication in arbuscular mycorrhizal symbiosis. *Curr. Opin. Plant Biol.* 16: 473-479.

Nealson, K.H. and Hastings, J.W. (1979). Bacterial bioluminescence: its control and ecological significance. *Microbiol. Rev.* 43:496–518.

Olah, B., Briere, C., Becard, G., Denarie, J. And Gough, C. (2005). Nod factors and a diffusible factor from arbuscular mycorrhizal fungi stimulate lateral root formation in *Medicago truncatula* via the DMI1/DMI2 signalling pathway. *Plant J.* 44: 195-207.

Orozco-Mosqueda Mdel, C., Macías-Rodríguez, L.I., Santoyo, G., Farías-Rodríguez, R., Valencia-Cantero, E. (2013a). *Medicago truncatula* increases its iron-uptake mechanisms in response to volatile organic compounds produced by *Sinorhizobium meliloti. Folia Microbiol.* 58:579-85.

Orozco-Mosqueda, M.C., Velázquez-Becerra, C., Macías-Rodríguez. L.I., Santoyo, G., Flores-Cortez, I., Alfaro-Cuevas, R. and Valencia-Cantero, E. (2013b). *Arthrobacter agilis* UMCV2 induces iron acquisition in *Medicago truncatula* (Strategy I plant) *in vitro* via dimethylhexadecylamine emission. *Plant Soil* 362:51–66.

Pineda, A., Zheng, S.J., van Loon, J.J.A., Pieterse, C.M.J. and Dicke, M. (2010). Helping plants to deal with insects: the role of beneficial soilborne microbes. *Trends Plant Sci.* 15:507–514.

Peñuelas, J., Asensio, D., Tholl, D., Wenke, K., Rosenkranz, M., Piechulla, B. et al. (2014). Biogenic volatile emissions from the soil. *Plant Cell Environ.* 37:1866–1891.

Pozo, M.J. and Azcon-Aguilar, C. (2007). Unraveling mycorrhiza-induced resistance. *Current Opin. Plant Biol.* 10: 393–398.

Rosta's, M, and Turlings, T.C.J. (2008). Induction of systemic acquired resistance in *Zea mays* also enhances the plant's attractiveness to parasitoids. *Biol. Control.* 46:178–186.

Rumbaugh, K.P., Griswold, J.A. and Hamood, A.N. (2000). The role of quorum sensing in the in vivo virulence of *Pseudomonas aeruginosa. Microbes Infect.* 2:1721–1731.

Schausberger, P., Peneder, S., Jurschik, S. and Hoffmann, D. (2012). Mycorrhiza changes plant volatiles to attract spider mite enemies. *Functional Ecology.* 26: 441–449.

Schloss, P.D. and Handelsman, J. (2004). Status of the Microbial Census. *Microbiol. Mol. Biol. Rev.* 68:686–691.

Schmidt, R., Cordovez, V., deBoer, W., Raaijmakers, J. and Garbeva, P. (2015). Volatile affairs in microbial interactions. *ISME J.* 9: 2329–2335.

Shimkets, L.J. (1999). Intercellular signaling during fruiting-body development of *Myxococcus xanthus. Annu. Rev. Microbiol.* 53:525–549.

Shtark, O.Y., Borisov, A.Y., Zhukov, V.A., Provorov, N.A. and Tikhonovich, I.A. (2010). Intimate associations of beneficial soil microbes with host plants. *Soil Microbial Sustainable Crop Prod.* 5: 119-196.

Singer, M.J. and Donald, N. Munns. (2006). Soils: an Introduction. Pearson Education Inc. New Jersey.

Smith, S.E. and Read, D.J. (2008). Mycorrhizal Symbiosis. Academic Press, London.

Son, S.H., Khan, Z., Kim, S.G. and Kim, Y.H. (2009). Plant growth-promoting rhizobacteria, Paenibacillus polymyxa and *Paenibacillus lentimorbus* suppress disease complex caused by root-knot nematode and fusarium wilt fungus. *J. Appl. Microbiol.* 107: 524–532.

Splivallo, R., Bossi, S., Maffei, M. and Bonfante, P. (2007). Discrimination of truffle fruiting body versus mycelial aromas by stir bar sorptive extraction. *Phytochemistry* 68: 2584–2598.

Splivallo, R., Fischer, U., Gobel, C., Feussner, I. and Karlovsky. P. (2009). Truffles regulate plant root morphogenesis via the production of auxin and ethylene. *Plant Physiology*. 150: 2018–2029.

Splivallo, R., Ottonello, S., Mello, A. and Karlovsky, P. (2011). Truffle volatiles: from chemical ecology to aroma biosynthesis. *New Phytol.* 189: 688–699.

Sprent, J.I. (2001). Nodulation in legumes. Kew Royal Botanical Gardens, Kew.

Steidle, A., Sigl, K., Schuhegger, R., Ihring, A., Schmid, M., Gantner, S., et al. (2001). Visualization of N-acylhomoserine lactone-mediated cell-cell communication between bacteria colonizing the tomato rhizosphere. *Appl. Environ. Microbiol.* 67: 5761–5770.

Sun, X., Bonfante, P. and Tang, M. (2015). Effect of volatiles versus exudates released by germinating spores of *Gigaspora margarita* on lateral root formation. *Plant Physiol. Biochem.* 97: 1-10.

Sylvia, D., Fuhrmann, J., Hartel, P., Zuberer, D. (2005). Principles and Applications of Soil Microbiology. Pearson Education Inc. New Jersey.

Strobel, G., Singh, S.K., Riyaz-Ul-Hassan, S., Mitchell, A.M., Geary, B. and Sears, J. (2011). An endophytic/pathogenic *Phoma* sp. from creosote bush producing biologically active volatile compounds having fuel potential. *FEMS Microbiol. Lett.* 320: 87-94.

Tan, R.X. and Zou, W.V. (2001). Endophytes: a rich source of functional metabolites. *Nat. Prod. Rep.* 18:448–459.

Thomson, N.R., Crow, M.A., McGowan, S.J., Cox, A. and Salmond, G.P.C. (2000). Biosynthesis of carbapenem antibiotic and prodigiosin pigment in *Serratia* is under quorum sensing control. *Mol. Microbiol.* 36:539–556.

Turlings, T.C.J., Loughrin, J.H., McCall, P.J., Rose, U.S.R., Lewis, W.J. and Tumlinson, J.H. (1995). How caterpillar-damaged plants protect themselves by attracting parasitic wasps. *PNAS.* 92: 4169–4174.

Unsicker, S.B., Kunert, G. and Gershenzon, J. (2009). Protective perfumes:the role of vegetative volatiles in plant defense against herbivores. *Curr Opin. Plant Biol.* 12: 479–485.

Uroz, S., Dessaux, Y., and Oger, P. (2009). Quorum sensing and quorum quenching: the yin and yang of bacterial communication. *Chembiochem* 10: 205-216.

Verheggen, F., Arnaud, L., Bartram, S., Gohy, M. and Haubruge, E. (2008). Aphid and plant volatiles induce oviposition in an aphidophagous hoverfly. *J. Chem. Ecol.* 34: 301–307.

von Bodman, S.B., Bauer, W.D. and Coplin, D.L. (2003). Quorum sensing in plant-pathogenic bacteria. *Annu. Rev. Phytopathol.* 41: 455–482.

Waters, C.M. and Bassler, B.L. (2005). Quorum sensing: cell-to cell communication in bacteria. *Annu. Rev. Cell. Dev. Biol.* 21:319–346.

Winter, T.R. and Rosta's, M. (2010). Nitrogen deficiency affects bottom-up cascade without disrupting indirect plant defense. *J. Chem. Ecol.* 36:642–651.

Zhang, Z. and Li, G. (2010). A review of advances and new developments in the analysis of biological volatile organic compounds. *Microchem. J.* 95:127–139.

In: Volatile Organic Compounds
Editor: Julian Patrick Moore

ISBN: 978-1-63485-370-5

Chapter 4

GASOLINE VEHICLES: THE PRIMARY CONTRIBUTOR TO AMBIENT VOLATILE ORGANIC COMPOUND (VOC) CONCENTRATIONS IN CALIFORNIA'S SOUTH COAST AIR BASIN

Yanbo Pang*, Paul Rieger and Mark Fuentes
California Air Resources Board, El Monte, CA, US

ABSTRACT

Volatile Organic Compounds (VOCs) play a significant role in the chemistry of air pollution at the local, regional, and global level by contributing to the formation of ozone and secondary organic aerosols (SOA). Some VOCs, including benzene, 1,3-butadiene, formaldehyde, and acetaldehyde are carcinogenic. Emissions from gasoline powered vehicles were known to be the main contributor to ambient VOC levels in all major cities of the world including the South Coast Air Basin (SoCAB) beginning in the 1960s. During the last five decades California led the world in taking action to reduce gasoline vehicle emissions.

Ambient VOC monitoring data have shown a significant decrease (>90%) in ambient VOC concentrations in the SoCAB since the 1960s.

* Corresponding author: Yanbo Pang. California Air Resources Board, 9528 Telstar Ave, El Monte, CA 91731. Email: ypang@arb.ca.gov.

Studies using data from ambient monitoring, highway tunnels, roadside measurements, and chassis dynamometers testing, showed that gasoline vehicle emission control required by regulatory agencies reduced emissions from gasoline powered vehicles more than 99 percent since the 1970s. This reduction of gasoline vehicle emissions is the main factor contributing to the decreases in ambient VOC concentrations in the SoCAB.

Given the large emission reductions from motor vehicles, it was anticipated that ambient VOCs would be dominated by sources other than tailpipe emissions and current emission inventories reflect this assessment. However, several recent inventory evaluation studies, based on the analysis of observed ambient data, showed that current emission inventories significantly underestimated gasoline powered vehicle emissions. One tunnel study indicated that the evaporative emissions of gasoline powered vehicles on hot days might be the major contributor to current ambient VOCs. However a recent study comparing trends in ambient VOC concentrations to trends in VOC emissions from gasoline vehicles tested on chassis dynamometers and to trends observed in tunnel studies showed that tailpipe emissions remained the main contributor to ambient VOCs in the SoCAB, in contradiction to the current inventory estimate and to studies suggesting that evaporative emissions explain the discrepancy. Control of tailpipe emissions from gasoline vehicles continues to be the best strategy for reducing ambient VOC levels and their harmful reaction products.

Keywords: gasoline powered vehicles, VOC, source contribution, regulation impacts

1. Introduction

Volatile Organic Compound (VOC) emissions from motor vehicles are known to degrade air quality. Most VOCs are precursors that catalyze the formation of tropospheric ozone and other oxidants, aromatic VOCs are precursors to the formation of secondary organic aerosols which increase the levels of ambient particulate matter (PM), and some VOCs are carcinogens (Koppman, 2008). Because of these air quality impacts, extensive measures have been taken by regulatory agencies to inventory and reduce tailpipe and evaporative emissions from motor vehicles (Parrish et al., 2002; O'Connor and Cross, 2006; Parrish et al., 2011; Warneke et al., 2012; Borbon et al., 2013; Propper et al., 2015; Pang et al. 2015). Various studies have been conducted to determine the effectiveness of these control strategies on emission reductions from vehicles and on ambient VOC concentration decreases (Warneke et al., 2012; Borbon et al., 2013; Fujita, et al. 2013; Propper et al., 2015; Pang et al. 2015).

Due to its very high population density (about 13 million people) and very inefficient ventilation, the South Coast Air Basin (SoCAB) also known as the Los Angeles (LA) basin has been plagued by the nation's most severe air quality problems since the 1950s (Haagen-Smit, 1952; Renzetti, 1956). The California Air Resources Board (CARB) emission inventory showed that mobile source emissions have been reduced by more than 90% in the SoCAB from 1975 to 2015 (CARB, 2007). Various studies also showed that the observed air quality improvement in the SoCAB during the last 40 years is primarily attributable to the emissions reductions from mobile sources (Parrish et al., 2011; Warneke et al., 2012; Pollack et al., 2013; Propper et al., 2015; Pang et al. 2015). However, there continues to be disagreement between inventory and apportionment studies regarding the relative contributions from source categories and the specific component of gasoline vehicle emissions that contributes to ambient VOCs. The relative contributions of tailpipe exhaust and non-tailpipe evaporative sources to ambient VOCs have been widely debated, especially in the last 20 years, even as large emission reductions have been achieved from regulations that tightened the emissions standards for gasoline powered vehicles (Pierson et al., 1999; Rubin et al. 2006; Fujita, et al. 2013; Pang et al. 2015). Understanding the relative contributions from tailpipe and evaporative emissions will advance the science of air pollution and facilitate the development of better control strategies in the future.

In this review, we summarize the current state of knowledge regarding the impact of emissions from gasoline powered vehicles on historical ambient VOC concentration trends, and on the current ambient urban atmosphere of the SoCAB. Furthermore, we discuss and evaluate recent studies that sought to determine the relative contributions of tailpipe and evaporative emissions in the SoCAB.

2. Definition and Measurement

VOCs are typically defined as carbon containing compounds having a vapor pressure greater than 10 Pa at 25°C, a boiling point of up to 260°C at atmospheric pressure, and 15 or less carbon atoms (Koppman, 2008). Other terms used to represent VOCs are hydrocarbons (HCs), reactive organic gases

(ROGs), and non-methane volatile organic compounds (NMVOCs). As a matter of practicality they are typically defined by the measurement techniques specified in the regulatory test procedures. California test procedures for Light Duty Gasoline Vehicles (LDGVs) specify the measurement of non-methane organic gases (NMOG), a surrogate for VOCs in vehicle exhaust. NMOG is determined utilizing three separate sampling and analysis techniques for quantifying C2-C12 hydrocarbons, C1-C6 carbonyl compounds and C1-C2 alcohols (CARB 2012). In monitoring ambient air, the most comprehensive database is for toxic VOCs which includes compounds such as benzene, toluene, formaldehyde, butadiene, etc. and typically utilizes Gas Chromatography/Mass Spectrometry (GC/MS) techniques for quantification. More comprehensive analyses of ambient VOCs, such as those carried out at Photochemical Assessment Monitoring Stations (PAMS), have been carried out in special projects, which are primarily aimed at understanding the ozone problem (Pang et al. 2015).

3. Sources of Ambient VOCs

The ambient air contains thousands of VOCs at concentrations in the range from parts per trillion to parts per billion and these compounds derive from dozens of major sources as well as from the photochemical reaction products of the primary source emissions. Ambient VOCs in urban areas arise primarily from anthropogenic sources such as motor vehicle exhaust, evaporation of gasoline vapors from cars, solvent usage, industrial processes, oil refining, gasoline storage and distribution, landfill emissions, food processing, and agricultural activities (Piccot, et al., 1992). Biogenic sources are also known to be significant contributors to ambient VOCs. Emissions from gasoline powered vehicles have been identified as the main contributor since the 1950s (Haagen-Smit, 1952; Haagen-Smit and Fox, 1954). Gasoline powered vehicles emit hundreds of VOCs that arise from incomplete combustion, unburned gasoline, and fuel evaporation. In general, gasoline powered vehicle emissions arise from both tailpipe exhaust and non-tailpipe evaporative sources. Tailpipe emissions include running exhaust and excess unburned fuel emissions associated with cold engine starting. Evaporative emissions include hot soak emissions that are driven by residual engine heat following vehicle operation, diurnal emissions associated with venting of fuel tank vapors as temperature increases during the day, running loss evaporative emissions that occur while vehicles are operating, and resting losses that result from gasoline permeation through rubber and plastic components of the fuel system (Rubin et al. 2006). Diesel powered vehicles can also be a significant contributor to the ambient atmosphere. However, inventory studies indicate that while Nitrogen Oxides (NOx) and PM emissions from diesel vehicles are substantial, the contribution to ambient VOCs from diesel vehicles is fairly minor compared to gasoline vehicles. VOC emissions and ambient air quality trends have been extensively studied over the years in the SoCAB for the purpose of source attribution and also for the purpose of determining the effectiveness of emission control strategies.

4. EMISSION INVENTORY

A VOC emission inventory combines bottom-up and top-down calculations for determining VOC emissions from various sources. The emission inventory plays an important role in the development of emission control strategies because policy makers use the emission inventory to determine the degree of cost effectiveness, in terms of tons per day pollutant removal, of potential control strategies. Emission inventories are calculated from emission factors which relate emissions of a given pollutant (for example, benzene mass per mile) and the level of the activity giving rise to that emission (for example, vehicle miles traveled). Emission inventories, by nature, are subject to significant levels of uncertainty because emission factors are developed from a limited number of measurements and activity data is based on assumptions about typical activity patterns. For example, vehicle emission factors are developed from a sample of vehicles driven on a dynamometer with a programmed duty cycle that may not be representative of the vehicle population. Furthermore, vehicle activity data are calculated based on a registered fleet that may not properly represent the activities of all vehicles on the road. Ambient measurements of VOCs provide a useful comparison with data calculated based on air quality models that use inventory data as inputs. Various studies using this comparison method in the SoCAB have been carried out (Parrish, 2006; Warneke et al., 2012; Fujita et al., 2013; Borbon et al., 2013). Most studies found substantial disagreement in the VOC/NOx ratio between the emission inventories and ambient measurements (Warneke et al., 2012; Fujita et al., 2013; Borbon et al., 2013).

5. Regulations on Emission Reductions and Their Impacts on Ambient VOC Concentrations

The CARB has led the world in the implementation of effective air pollution controls ever since the introduction of the nation's first motor vehicle emission standards in 1966. Early standards led to power train modifications such as positive crankcase ventilation, the introduction of air pumps, fuel metering and engine timing changes (O'Connor and Cross, 2006). In the seventies, focus shifted to exhaust after-treatment with the introduction of catalyst technologies to reduce VOCs and NOx including oxidizing unburned fuel. In the 1980s, the recognition that a small number of high emitting vehicles could contribute the largest fraction of emissions prompted the improvement of SMOG Check programs and the promulgation of on board diagnostic regulations. Evaporative emissions both from the vehicle and from the refueling process were also addressed during this time with the implementation of the vapor recovery regulations and improved evaporative testing protocols (O'Connor and Cross, 2006). In an effort to assure that future air quality standards for ozone were met, the CARB required auto manufacturers to meet the first standards to control smog forming hydrocarbon and NOx emissions. In the 90s, the Low-Emission Vehicle (LEV) regulations targeted the ozone forming potential of emissions as determined from the aggregated incremental reactivity of all VOCs in exhaust which were defined in the NMOG test procedures related to the regulations (CARB 2012). An important adjunct to the LEV regulations was California's Clean Fuels program, initiated in the 1990s, which led to the reformulation of gasoline and diesel fuels (O'Connor and Cross, 2006). The large reduction in sulfur content specified in the regulations assured that catalysts required by the LEV regulations would remain effective in reducing ozone precursors, while the regulated specifications for benzene, aromatic, vapor pressure, distillation points and olefins assured that toxic compounds and reactive hydrocarbons would also be reduced. The clean fuel regulations coupled with the LEV regulations have resulted in a 99 percent reduction of pollutants from modern vehicles relative to their predecessors thirty years ago (CARB, 2016).

The effectiveness of these mobile source regulations have been evaluated by studies using ambient VOC measurement data (Warneke et al., 2012; Pollack et al., 2013; Propper et al., 2015; Pang et al., 2015). These studies, discussed in the following paragraphs, concluded that the CARB mobile source emission control strategies played an essential role in the large decreasing trend in ambient VOC concentrations (>90%) in the SoCAB.

By reviewing and combining the results from continuous monitoring stations with the results from multiple special studies, Pollack et al. (2013) calculated the five decade (1960-2010) ambient concentration trends for ozone, its precursors and other secondary oxidation products to access impacts of regulations on air pollutant concentration decreases. Their study was based on data from short-term field studies, aircraft measurements, and near-tailpipe measurements. Several short-term ground-based field studies have been conducted at selected locations within the SoCAB since 1960. Basin-wide measurements from instrumented research aircraft began in the 1970s (Husart et al., 1977), and near-tailpipe measurements from mobile roadside monitors began in the early 1990s (Lawson et al., 1990; Beaton et al., 1995; Gertler et al., 1999; Bishop and Stedman, 2008). Long-term trends in ozone and emissions of its precursors in the SoCAB have been extensively studied using the data collected in these experiments (Ban-Weiss et al., 2008; Bishop and Stedman, 2008; Dallmann and Harley, 2010; Fortin et al., 2005; Fujita et al., 2003; Fujita et al., 2013; Grosjean, 2003; Harley et al., 2005; McDonald et al., 2012; Parrish et al., 2002; Parrish et al., 2011; Warneke et al., 2012).

The study by Pollack et al. (2013) showed an average rate of decrease of $7.3 \pm 0.7\%\ yr^{-1}$ for anthropogenic VOCs, corresponding to a decrease in average abundances of a factor of 44 over 50 years, and was determined using data from the Azusa and Upland South Coast Air Quality Monitoring District (SCAQMD) stations and the ground-based field measurements since 1960. Pollack et al. (2013) also indicated that the annual average decrease in ROG of $6.2 \pm 0.4\%\ yr^{-1}$ between 1975 and 2010 solely from SoCAB on-road mobile sources in the CARB inventory is in better agreement with ambient observations than is the annual average decrease of $4.4 \pm 0.4\%\ yr^{-1}$ determined from all sources in the CARB inventory.

Warneke et al. (2012), reviewed $\Delta CO/CO2$ and $\Delta VOC/CO$ ratios during the last five decades and found that VOC and CO concentrations decreased by almost two orders of magnitude from the 1960s to 2010 and VOC mixing ratios decreased at an annual average rate of 7.5% during this period. Propper et al. (2015) reported that toxic VOC concentrations decreased by 80% from 1990 to 2009 mainly due to vehicle emission reductions in the SoCAB. Pang et al. (2015) reported that annual median concentrations for most ambient VOCs decreased 40% from 1999 to 2009 in the SoCAB, based on data from the PAMS. These studies all indicated that ambient VOC concentration decreases were due to tailpipe emission reductions from gasoline powered vehicles. Parrish et al. (2009) concluded that US mobile source emission source control strategies played an important role in the ambient VOC concentration decreases in US mega cities. Emission inventory data also confirm this conclusion.

6. UNDERESTIMATION OF VOC EMISSIONS BY EMISSION INVENTORY

Current emission inventories show large mobile source emission reductions during the last four decades which can be attributed to control strategies implemented by European and US environmental policy makers including the CARB (Borbon et al., 2013). The current emission inventory data for the SoCAB indicates that mobile source categories, in particular gasoline powered vehicles, are no longer the dominant ambient VOC source (CARB, 2007), in Figure 1. Diesel powered vehicles only contributed less than 2% from 1975 to 2015, as shown in Figure 1. In 1975, gasoline powered vehicles and all mobile sources contributed 55% and 63% of VOCs respectively in the SoCAB. In 2015, gasoline powered vehicles only contributed 17% of ambient VOCs. According to the inventory, area wide sources mainly from solvent use became the new dominant source and contributed 27% of ambient VOCs. However, recent ambient VOC measurement studies showed that the current emission inventory underestimated the contributions from mobile sources, especially from gasoline powered vehicles (Parrish, 2006; Warneke et al., 2012; Fujita et al., 2013; Borbon et al., 2013).

Parrish (2006) compared US Environmental Protection Agency (EPA) estimates of on-road vehicle emissions to ambient measurements and to a fuel-based emission inventory from 1990 to 2005 including LA basin. His study showed that the most recent EPA emissions estimate accurately captures the rapid decrease in carbon monoxide (CO) and VOC emissions. However, the ratios of two specific ambient VOC species (acetylene and benzene) suggested that the inventory speciation of VOCs was inaccurate by factors of 3-4.

Warneke et al. (2012) reviewed data from 15 intensive ambient VOC measurement campaigns and found that the mixing ratios of CO and VOC have decreased by almost two orders of magnitude in the past fifty years at an average annual rate of almost 7.5%.

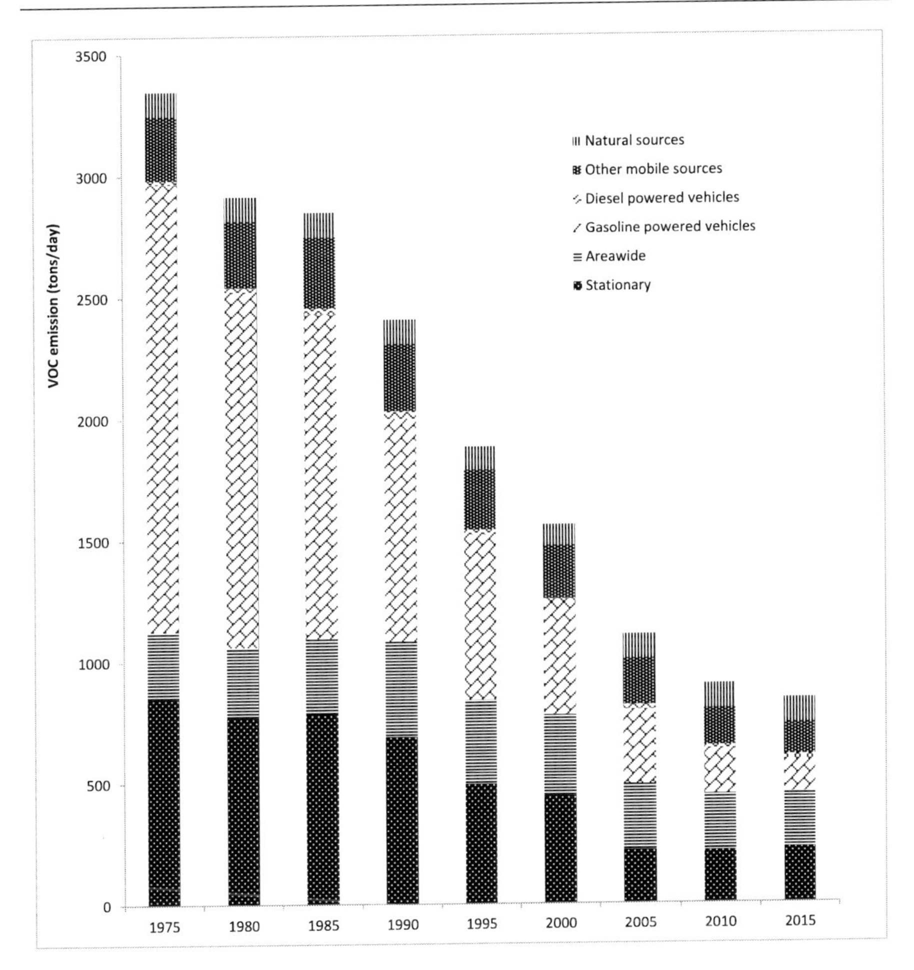

Figure 1. Emission inventory from 1975 to 2015 in the SoCAB.

Warneke et al. (2012) also reported that VOC/CO ratios have been remarkably constant for the period 1960 to 2010. This suggested that the main VOC source in the SoCAB remained gasoline powered vehicle emissions.

Fujita et al., 2013 analyzed SoCAB ambient monitoring data from 1980 to 2010. By comparing VOC/NOx ratios, Fujita et al. found that the average ambient NMOC/NOx ratios increased from 3.9 in 1997 to 5.4 in 2009. At the same time, the average of the emission inventory ROG/NOx ratios for the corresponding locations decreased from 3.7 in 1997 to 2.8 in 2009. With these diverging trends in the ambient and emission inventory VOC/NOx ratios, the ambient ratios were again higher by about a factor of 2 than the emission inventory ratios in 2009 and suggested a factor of 2 underestimation of VOC emissions relative to NOx. Fujita et al. (2013) suggested that the evaporative emissions of gasoline vehicles on hot days may be the major contributor to this underestimation.

Borbon et al. (2013) used ground-based and airborne VOC measurements in Los Angeles, California, and Paris, France, during the Research at the Nexus of Air Quality and Climate Change (CalNex) and Megacities: Emissions, Urban, Regional and Global Atmospheric Pollution and Climate Effects, and Integrated Tools for Assessment and Mitigation (MEGAPOLI) campaigns, respectively, to evaluate emission inventory estimates. Emission ratios for alkenes, alkanes, and benzene were found to be fairly similar between Los Angeles and Paris, whereas the emission ratios for C7-C9 aromatics in Paris were higher than in Los Angeles and other French and European Union urban areas by a factor of 2-3. The results suggested that the emissions of gasoline-powered vehicles still dominated the hydrocarbon distribution in northern mid-latitude urban areas, in contradiction to the data in emission inventories which pointed to the use of solvents as the dominant source.

In summary, these recent studies comparing ambient measurement data with emission inventory estimates clearly showed that current emission inventories underestimated VOC emissions from gasoline-powered vehicles. However, emissions from gasoline powered vehicles include both tailpipe and evaporative emissions. The remaining question is which one, tailpipe emission or evaporative emission, is the main contributor for VOC emissions from gasoline powered vehicles.

7. Tailpipe Emissions from Gasoline Powered Vehicles, the Main Contributor of Ambient VOCs

Fujita et al. (2012) conducted an on-road mobile source emission study at a traffic tunnel in Van Nuys, California, in August 2010 to measure fleet-averaged, fuel-based emission factors while simultaneously carrying out remote sensing device (RSD) measurements near the tunnel. The tunnel and RSD fleet-averaged emission factors were compared with the corresponding modeled factors calculated using US Environmental Protection Agency's (EPA's) MOVES2010a (Motor Vehicle Emissions Simulator) and MOBILE6.2 mobile source emission models, and California Air Resources

Board's (CARB's) EMFAC2007 (EMission FACtors). They concluded that evaporative emissions of gasoline powered vehicles on hot days may be the major contributor to ambient VOCs.

In this study, Fujita et al. (2012) reported that the non-methane hydrocarbon (NMHC) emission factors (specifically the running evaporative emissions) predicted by MOVES were insensitive to ambient temperature as compared with the tunnel measurements and compared with the MOBILE- and EMFAC-predicted emission factors, resulting in underestimation of the measured NMHC/NOx ratios at higher ambient temperatures. Tunnel measurements showed that the measured NMHC emission factors were 2.0 and 3.5 times higher at 85-95 and 95-105°F, respectively, compared to emission factors measured at ambient temperatures of 65-75°F. The ratios of modeled evaporative to exhaust emission factors at 104°F relative to 65°F are 4.0, 1.7, and 1.1 for EMFAC, MOBILE, and MOVES, respectively. The corresponding ratio of the estimated evaporative emission factor from the tunnel measurements at ambient temperatures of 101-102°F relative to 70-72°F was about 6. Fujita et al. (2012) concluded that the MOVES model predicts running evaporative emissions that are in reasonable agreement with the estimates from measurements at lower ambient temperatures, but, because the model is inadequately sensitive to ambient temperature, underestimates the emissions at higher temperature. The MOBILE model was found to predict the highest evaporative emission factors, but showed less temperature sensitivity than indicated by the tunnel measurements. The apparent temperature sensitivity observed in the tunnel measurements is best replicated by EMFAC, though it too underestimated NMHC emissions at high ambient temperatures.

Fujita et al. (2012) also used source VOC markers to evaluate impacts of temperature on source strengths. Ambient temperature generally has a greater impact on the evaporative emissions of the more volatile hydrocarbons, such as isopentane and n-butane, in gasoline (Rubin et al., 2006). On the other hand, acetylene, not found in gasoline, is a marker for vehicle combustion or tailpipe emissions and is relatively insensitive to temperature changes (Pang et al. 2014 and 2015). Fujita et al. (2012) found that the ratios of the measured emission factors at high and low ambient temperatures were 0.9 for acetylene and 8.1 for n-butane. They concluded that the models, especially MOVES, may not fully account for running evaporative emissions at higher ambient temperature based on these tunnel VOC speciation data.

Fujita et al. (2013) used ambient VOC measurement data to confirm the result of their tunnel study. They studied the combined mean of the diurnal variations in the ratios of NMOC/NOx and isopentane/acetylene (ratio of a component of gasoline to a major combustion product in vehicle exhaust) at Los Angeles N. Main and Azusa during summer 2009. Fujita et al. (2013) segregated the data into two ambient temperature bins: days with temperatures equal to and above 85°F and days below 85°F. The study reported that the midday NMOC/NOx and isopentane/acetylene ratios were both about 30% higher during the warmer days. The effect of ambient temperature on NMOC/NOx ratios and results of the 2010 Van Nuys Tunnel Study are generally consistent with an ambient source apportionment study that estimated a 6.5%-2.5% increase in the contributions of evaporative emissions from motor vehicles per degree Celsius increase in maximum temperature (Rubin et al., 2006). Fujita et al. (2013) concluded that the emission inventory VOC/NOx ratio underestimation was caused by the underestimation, of evaporative emissions during hot days by motor vehicle emissions models.

Pang et al. (2015) re-analyzed these Fujita et al. (2012 and 2013) studies and raised several questions regarding their results. Fujita et al. (2012) found that 20°F temperature changes doubled emission rates of evaporative emission markers (isopentane) and had no impact on rates of the tailpipe exhaust markers (acetylene). However, the low temperature emission factors in the study were determined by only two weekend day morning measurements and the high temperature factors were from two weekday afternoon measurements. However, emissions in a traffic tunnel are quite different from weekend to weekdays as well as morning to afternoon. The temperature effect would be more accurately determined if the emission factors were calculated using tunnel data of similar (day of week) time periods with different temperatures. Furthermore, more data points are needed for a conclusive determination of the temperature effect.

Pang et al. (2015) found that the ambient temperature did not have a large effect on VOC ratios. Acetylene and isopentane ratios remained relatively invariant in the diurnal data from the PAMS sites. Figure 2-4 shows that from 1998 to 2008 the changes in hourly benzene-concentration normalized ratios were small - from 0.7 to 1 (the minimum to the maximum). The relatively small diurnal change in ratios suggests that one dominant source contributed most of the ambient VOCs on a daily basis, especially since other VOC sources have very different VOC profiles (Watson et al., 2001; EPA, 2012). This relatively small diurnal change also indicates that this dominant source, unlike evaporative emissions, is not sensitive to temperature changes because temperatures have higher impacts on ratios of compounds related to evaporative emissions, such as butane and isopenatne, etc. (Rubin et al., 2006).

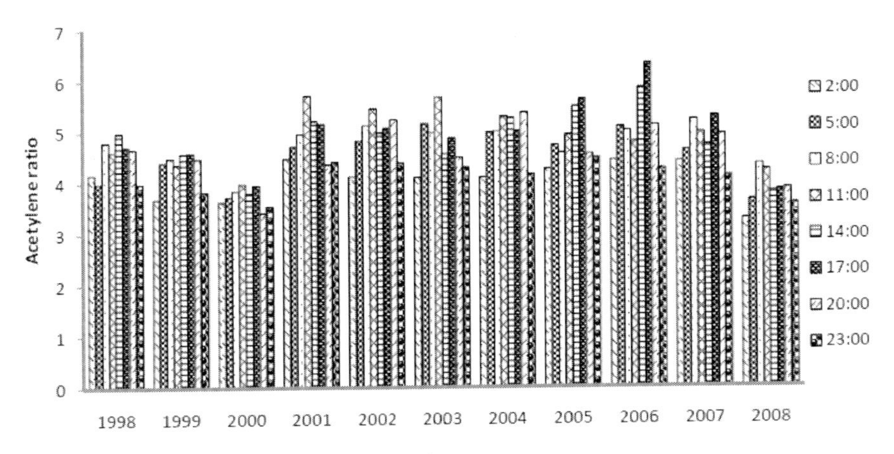

Figure 2. Trends in hourly benzene concentration normalized acetylene ratios from 1998 to 2008 at Upland, California.

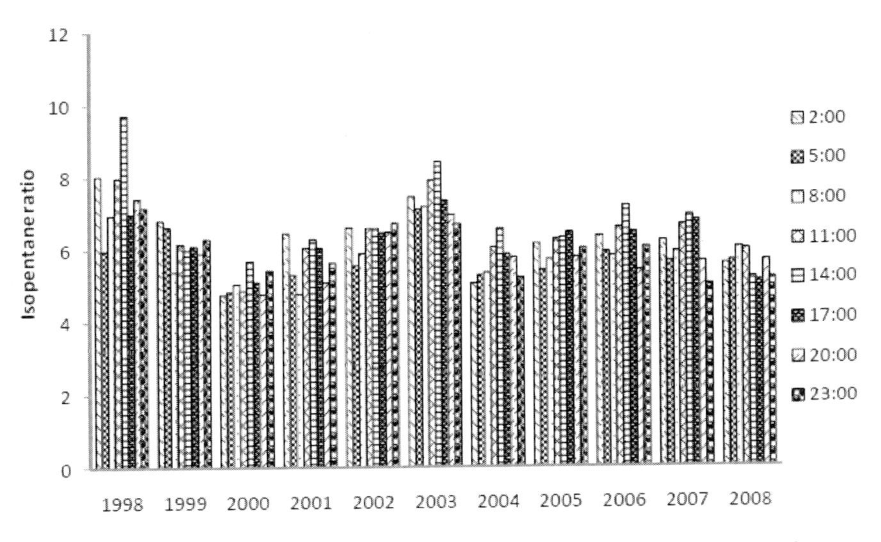

Figure 3. Trends in hourly benzene concentration normalized isopentane ratios from 1998 to 2008 at Upland, California.

Figure 3 shows that, from 1998-2008, the lowest isopentane ratio appeared at night (23:00, 2:00, and 5:00), except in 2008 when it appeared at 17:00, and in 2006 and 2000 when it appeared at 20:00. The ratio of the highest isopentane/benzene ratios to the lowest ratios for 1998 to 2008 ranged from 1.38 to 1.18 (except 1.62 for 1998) even though the average diurnal temperature change in the Los Angeles area was about 20°F (Los Angeles Almanac 2015).

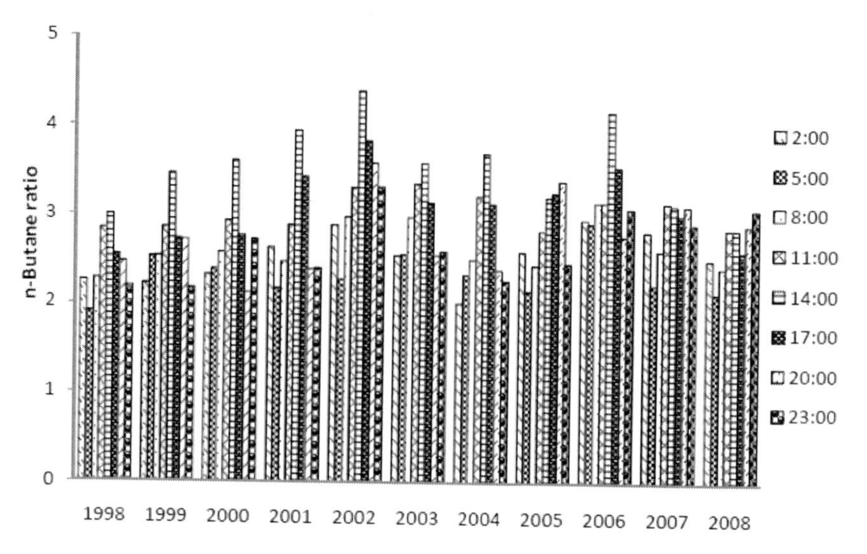

Figure 4. Trends in hourly benzene concentration normalized n-butane ratios from 1998 to 2008 at Upland, California.

However, according to the Fujita calculation this ratio should be greater than 2.0 (Fujita et al., 2012). In addition, Figure 2 shows that acetylene, the marker for tailpipe exhaust emissions, had the same diurnal pattern as isopentane. The ratios of the highest acetylene ratios/the lowest were 1.18 to 1.38 from 1998 to 2008 except 1.48 for 2007. The data in this study showed that ambient temperature did not have differing impacts on the marker compounds for tailpipe and evaporative emissions. N-butane data in Figure 4 also showed the same pattern as acetylene data.

Pang et al. (2015) also applied the Fujita et al. (2013) method to analyze trends in isopentane/acetylene, isopentane/2-methylpentane, and n-butane/ isopentane ratios from 1998 to 2009 using PAMS data at the Azusa, California site. The data are shown in Figures 5-7. The isopentane/acetylene ratios for hot days were generally higher than those for cold days by 15-55% in 2009 as shown in Figure 5. But, no consistent trends in the diurnal patterns were found for data from 1998 to 2009. The ratio differences for hot and cold days in 2009 are similar to the result of the Fujita et al. (2013) study. However, isopentane/ acetylene ratios for hot days were not always higher than those for cold days, especially for the years 2000, 2004 and 2007, from 1998 to 2009 (Figure 5). The importance of contributions from evaporative emissions during hot days cannot be readily inferred by the method proposed by the Fujita et al. study (2013). Comparisons of ratios of isopentane/2-methylpentane and ratios of n-butane/isopentane for hot and cold days also show no indication that those more volatile organic compound emissions increased more significantly during hot days, as shown in Figure 6 and Figure 7. Pang et al. (2015) concluded that tailpipe emissions remained a dominant contributor to ambient VOCs relative to evaporative emissions from gasoline powered vehicles.

In their study, Pang et al. (2015) compared trends in ambient concentrations of VOCs in the SoCAB to trends in VOC emissions from gasoline powered vehicles tested on chassis dynamometers and to emissions trends observed in tunnel studies during the same period. This study showed that tailpipe emissions from gasoline powered vehicles remained the dominant source of ambient VOCs in the SoCAB in 2009 even though tailpipe emissions decreased significantly.

Figure 8 shows that benzene concentration normalized ratios for most compounds, such as acetylene, ethene, >C5 alkanes and aromatics including toluene and o-xylene, remained stable from 1999 to 2009 even though most compound concentrations decreased more than 50% (Pang et al., 2015). These trends suggest that the primary sources of most VOCs in the summer remained the same from 1999 to 2009 in the SoCAB. The doubling of ethane and propane ratios is primarily due to the decrease in benzene concentrations (~50%) from 1999 to 2009, in Figure 8. These two compounds are known to come mainly from sources other than mobile sources (OEHHA, 2006; Navazo et al., 2008). The ambient 2,2,4-trimethylpentane ratios started to increase in 2003 consistent with the increase in tailpipe exhaust emission rate ratios observed in the 2003 fleet of in-use gasoline powered vehicles and was attributable to the California Phase 3 gasoline regulation which prompted an increasing fraction of 2,2,4-trimethylpentane in gasoline (Pang et al., 2014).

Figures 9 and 10 show the comparison of the time-series of gasoline powered vehicle emissions in the Caldecott tunnel studies and the trends of gasoline powered vehicle fleet emissions from dynamometer testing (Harley, 2004; Harley, 2009, Pang, et al. 2014). Benzene mole concentration normalized ratios for gasoline powered vehicle emissions in the Caldecott tunnel (Harley, 2004; Harley, 2009) and for fleet average tailpipe and evaporative emissions (Pang et al., 2014) for 1999 and 2005 are shown in Figure 9 and Figure 10. Data in Figure 9 and 10 indicate that the VOC patterns in the tunnel emissions are more similar to tailpipe than to evaporative emissions for both the 1999 and 2003 fleets reported in the Pang et al. (2014) study. Linear correlation coefficients between ratios of tailpipe exhaust and tunnel for 1999 and 2005 were 0.9 and 0.8, respectively. Correlations between evaporative and tunnel profiles for 1999 and 2005 were poor (correlation coefficient is 0.03 for 1999 and 0.22 for 2005). Therefore, tailpipe emissions were likely the main contributor to gasoline vehicle emissions in the Caldecott tunnel for 1999 and 2005. This is consistent with results from studies at the Fort McHenry and the Tuscarora mountain tunnels where 85% of observed LDGV NMHC emissions were determined to be from tailpipe exhaust emissions (Gertler et al., 1996).

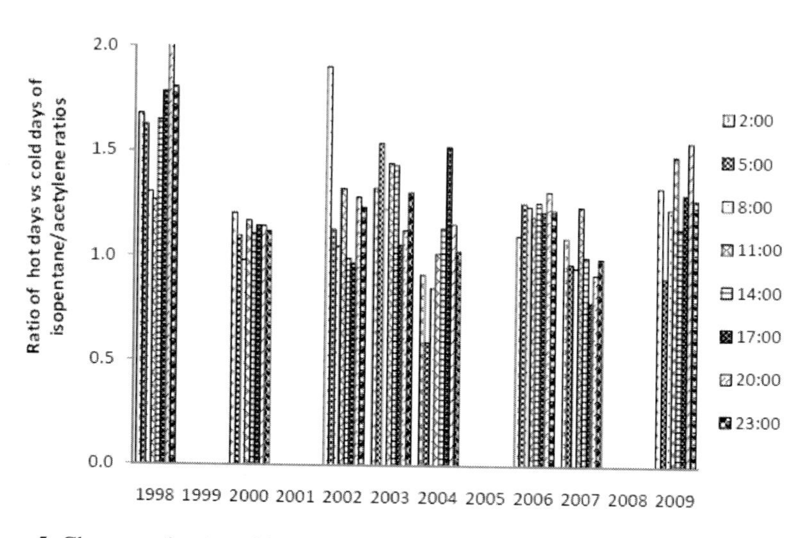

Figure 5. Changes of ratios of hot days vs cold days for isopentane/acetylene ratios from 1998 to 2009.

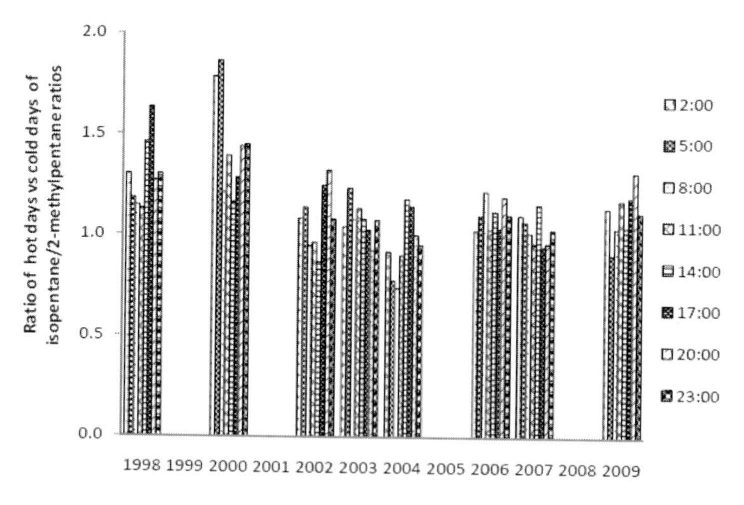

Figure 6. Changes of ratios of hot days vs cold days for isopentane/2-methylpentane ratios from 1998 to 2009.

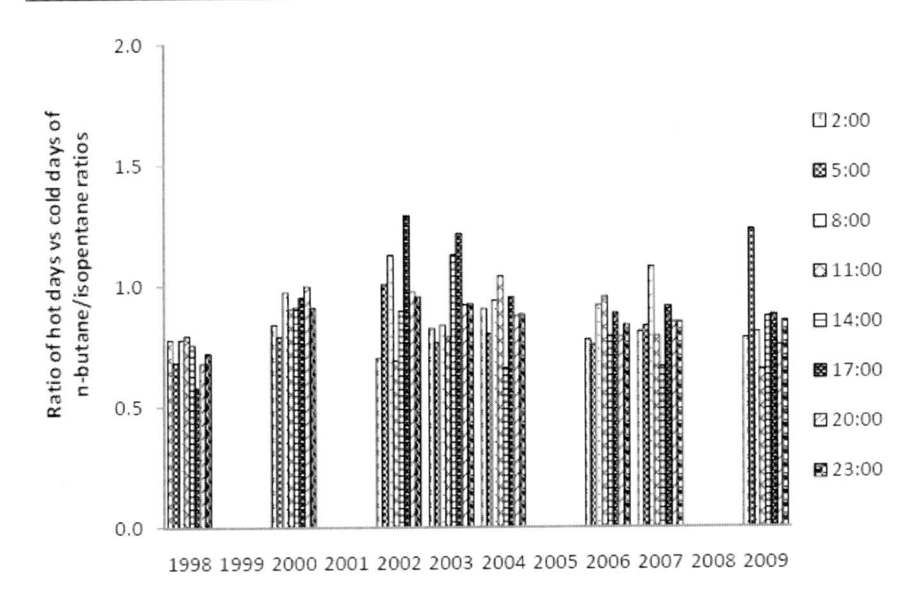

Figure 7. Changes of ratios of hot days vs cold days for n-butane/isopentane ratios from 1998 to 2009.

Data for compounds with large ratio differences between tailpipe and evaporative emissions also indicate that tailpipe emissions were the dominant contributor to tunnel emissions, in Figure 9 and 10. For example, isobutane and n-butane ratios for evaporative emissions are 10-100 times higher than both ratios for tailpipe emissions. Chemical Mass Balance (CMB) emission source profiles also show similar differences (Watson et al., 2001). These two compound ratios in the Caldecott Tunnel emissions, in Figure 9 and 10, were close to those of tailpipe emissions for both the 1999 fleet and the 2003 fleet.

Pang et al. (2014) indicated that VOC patterns for tailpipe emissions were stable while patterns for evaporative emissions changed dramatically from the 1999 fleet to the 2003 fleet. Changes in VOC patterns in tunnel studies were similar to those observed in tailpipe emissions. The weight percentages of benzene in total NMHC in tunnel studies were close to the weight percentages in tailpipe exhaust emissions (Figure 9 and 10). Compound ratios in tailpipe emissions and Caldecott Tunnel emissions in Figure 9 and 10, especially for aromatics, are also close to those determined in European tunnels (Stemmler et al., 2005; Petrea, 2007). This similarity in global tunnel results indicates that tailpipe emissions not evaporative emissions are dominant in gasoline powered vehicle emissions in all tunnel studies.

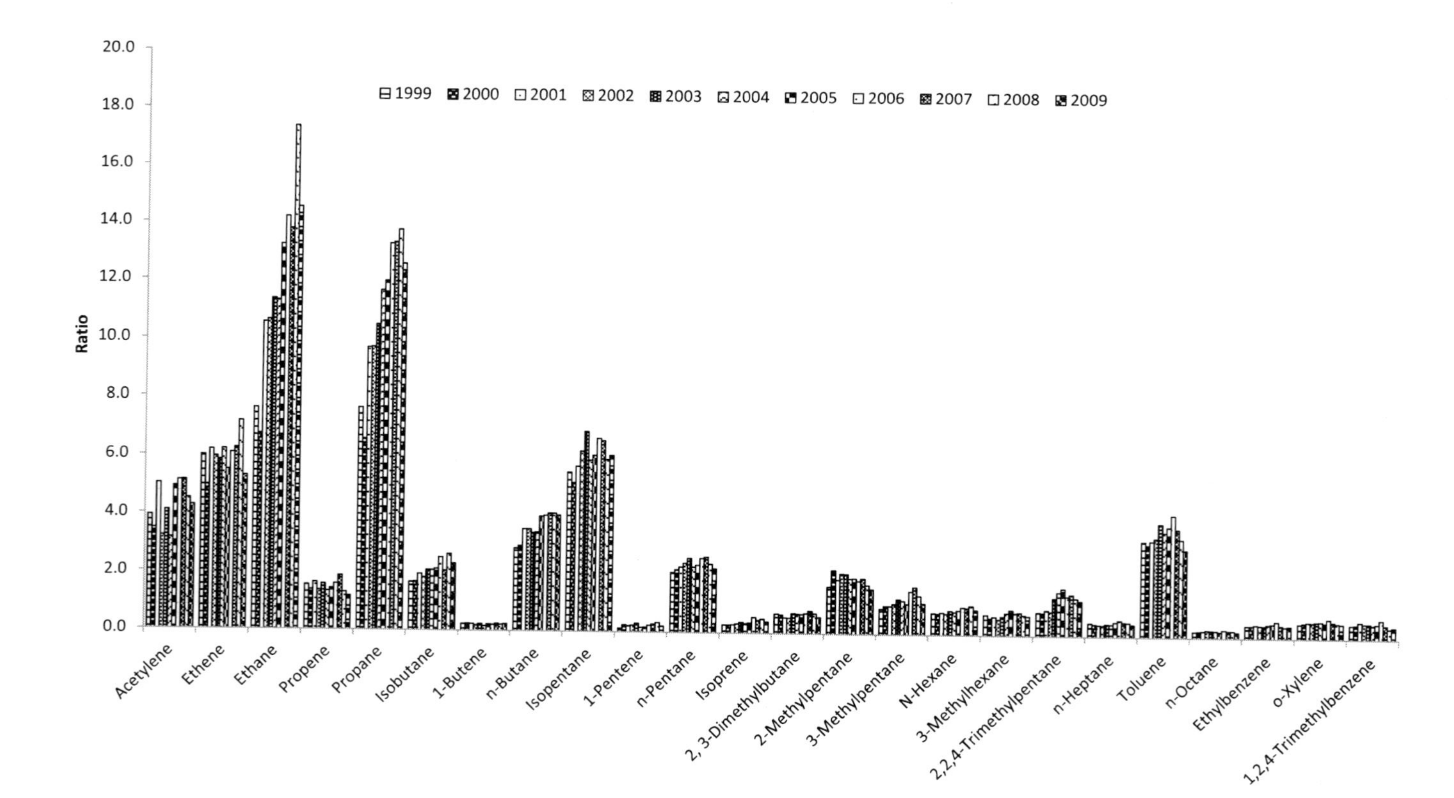

Figure 8. Trends in annual VOC benzene concentration normalized ratios from 1999 to 2008 in the SoCAB.

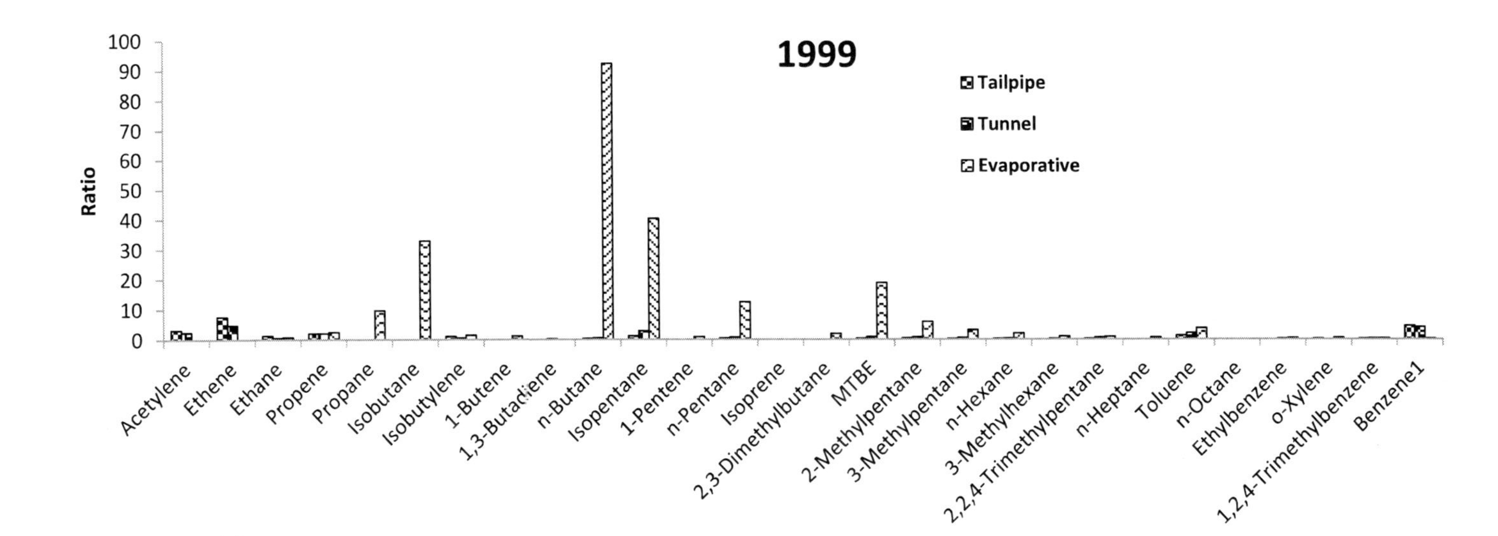

1: benzene weight percentage.

Figure 9. Comparison of benzene-normalized ratios of tunnel emission factors to tailpipe and evaporative fleet average emissions for 1999.

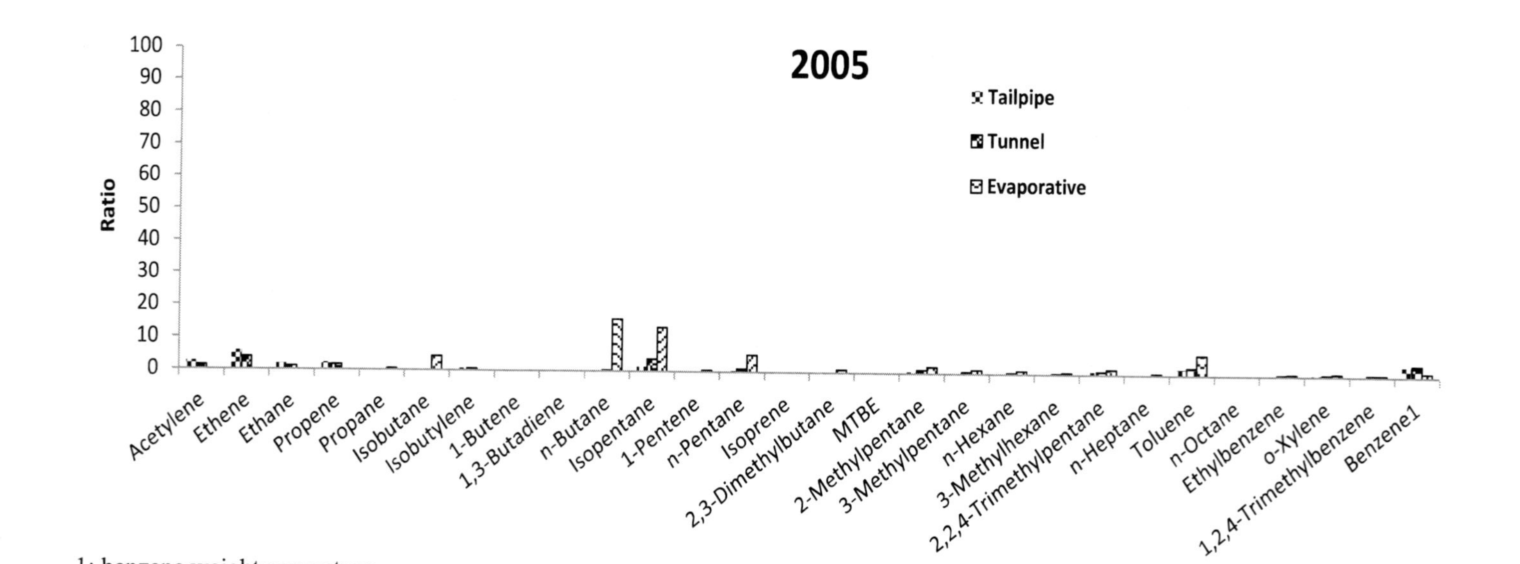

1: benzene weight percentage.

Figure 10. Comparison of benzene-normalized ratios of tunnel emission factors to tailpipe and evaporative fleet average emissions for 2005.

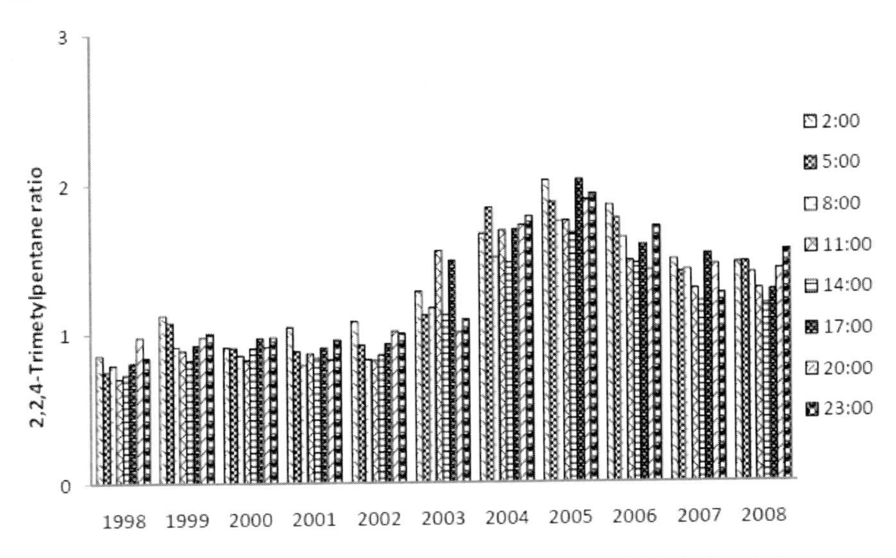

Figure 11. Trends in hourly benzene concentration normalized 2,2,4 trimethylpentane ratios from 1998 to 2008 at Upland, California.

Pang et al. (2015) indicated that the data at two individual sites, Azusa and Upland, in the SoCAB showed the same result. Those trends in ambient VOC data are more similar to tailpipe exhaust emission trends than evaporative emission trends (Pang et al. 2014). The similarities in VOC trends for the ambient atmosphere, traffic tunnels and fleets tested on dynamometers suggest that tailpipe emissions remained the dominant source for ambient VOCs in the SoCAB.

The diurnal patterns of ambient VOC concentrations depend on temporal variations in the strengths of the emission sources, as well as on photochemical activity and meteorological conditions. The diurnal pattern is known to reflect the impacts of mobile sources during the day. For example, the difference between 2:00 and 8:00 in the morning is an indicator of morning traffic activities. Figures 2, 3 and 11 show that the changes in hourly benzene-concentration normalized ratios from 1998 to 2008 were small for most compounds. The relatively small diurnal changes in ratios suggest that one dominant source contributed most of the ambient VOCs on a daily basis and this dominant source, unlike evaporative emissions, is not sensitive to temperature changes (Rubin et al., 2006).

Pang et al. (2015) indicated that changes in the annual median compound ratios between different years from 1999 to 2009 were even smaller than the changes in hourly compound ratios for any given year except for ethane, propane, isoprene and 2,2,4-trimethylpentane even though most VOC concentrations decreased more than 40% during these years. These small changes in annual hourly ratios confirm that the main sources of most ambient VOCs in 2008 remained the same as those observed 10 years earlier (gasoline vehicles were the main source of ambient VOCs). Diurnal patterns for the 2,2,4-trimethylpentane ratio during the gasoline Phase 3 regulation implementation period (2003-2004) were constant even though ratios increased dramatically, in Figure 11. The CARB gasoline Phase 3 regulation caused the 2,2,4-trimethylpentane ratio to increase for gasoline powered vehicle emissions (Pang et al., 2014). This correspondence between ambient 2,2,4-trimethylpentane trends and gasoline powered vehicle 2,2,4-trimethylpentane emission trends further confirms the dominant impact of the contribution of gasoline powered vehicles to ambient VOCs.

The current inventory model underestimation of VOC emissions from mobile sources, has been reported in several recent studies of ambient trends in the SoCAB (Parrish, 2006; Warneke et al., 2012; Fujita et al., 2013; Borbon et al., 2013. This new study showed that tailpipe emissions remained a bigger contributor to ambient VOCs than evaporative emissions from gasoline powered vehicles (Pang et al., 2015). This conclusion is inconsistent with data in current emission inventories. However, the reasons for the underestimation of VOC emission by inventories are still unknown. Pang et al. (2015) indicated that high emitters (malfunctioning vehicles), off-cycle emissions, higher tailpipe emissions on hot days or other (unknown) sources may be responsible for this discrepancy.

Rubin et al. (2006) used a chemical mass balance approach to determine the relative contributions of evaporative versus tailpipe sources to motor vehicle volatile organic compound (VOC) emissions. Continuous VOC measurements were recorded with time resolution of 45 min over an 8-week period from 18 July through 15 September 2001 using online gas chromatographic methods (Millet et al., 2005). Based on these highly time resolved ambient VOC concentration data, this study found that evaporative emission contributions ranged from 7 to 29% of total vehicle-related VOC depending on time of day and day of week, with a mean daytime contribution of 17.0 ± 0.9% (mean ± 95% CI). It indicated that tailpipe emissions from gasoline powered vehicles were the dominant source for VOC emission from vehicles. Rubin et al. (2006) also concluded that emission inventory models underestimate the importance of the contribution from tailpipe emissions of gasoline powered vehicles. This study provided the basic understanding of temperature impacts on evaporative emissions, which is used by Fujita et al.

(2013) and Pang et al. (2015). However, this study was conducted more than 15 years ago.

The Rubin et al. (2006), Fujita et al. (2013) and Pang et al. (2015) studies provided different conclusions regarding the relative contributions of tailpipe vs evaporative emissions from gasoline powered vehicles even though most recent studies agreed that current models underestimated VOC emissions from gasoline powered vehicles (Parrish, 2006; Warneke et al., 2012; Fujita et al., 2013; Borbon et al., 2013; Pang et al., 2015). Therefore, more research is still needed to understand the contributions of gasoline powered vehicles to ambient VOCs.

CONCLUSION

Emissions from gasoline powered vehicles are known to be a major contributor to the ambient atmosphere in cities around the world and control measures have been taken to reduce these emissions since the 1960s. As control measures were taken during the last five decades, ambient VOC measurement and roadside studies suggested that the emissions inventory still underrepresented VOC emissions from gasoline powered vehicles. Adjustments to emissions models have improved emission inventories and reduced the VOC/NOx ratio discrepancy for a period of time but more recently the discrepancy has widened again.

Recent ambient VOC measurement studies showed that gasoline vehicle emissions were still the major VOC contributor to the ambient air in contradiction to current emission inventories in the SoCAB. Even though all these studies agreed that current emission inventories underestimated emissions from gasoline powered vehicles, it is still unclear which component of vehicle emissions is responsible for this underestimation. Based on speciated hydrocarbon measurements in traffic tunnels and the ambient air, some investigators attributed the discrepancy to an underestimation of evaporative emissions from gasoline powered vehicles during hot days. However, a more comprehensive comparison of ambient ratio trends, tunnel studies and diurnal data suggested that tailpipe emissions from gasoline powered vehicles remained the dominant source of ambient VOCs in the SoCAB. Given the importance of emissions inventories to environmental policy it is important that the current discrepancy between inventory and ambient measurements is resolved.

Disclaimer: The statements and opinions expressed in this paper are solely the authors' and do not represent the official position of the California Air Resources Board. This paper has not been subjected to CARB's peer and policy review, and official endorsements should not be inferred. The Air Resources Board is a department of the California Environmental Protection Agency. CARB does not endorse nor promote any of the opinions or suggestions inferred by the use of the trademarks that appear in this paper.

REFERENCES

Ban-Weiss, G.A., McLaughlin, J.P., Harley, R.A., Lunden, M.M., Kirchstetter, T.W., Kean, A.J., Strawa, A.W., Stevenson, E.D. and Kendall, G.R., 2008. Long-term changes in emissions of nitrogen oxides and particulate matter from on-road gasoline and diesel vehicles. *Atmospheric Environment*, 42 (2), pp. 220-232.

Beaton, S.P., Bishop, G.A., Zhang, Y. and Ashbaugh, L.L., 1995. On-road vehicle emissions: regulations, costs, and benefits. *Science*, 268(5213), p. 991.

Bishop, G.A. and Stedman, D.H., 2008. A decade of on-road emissions measurements. *Environmental Science and Technology*, 42(5), pp. 1651-1656.

Borbon, A., Gilman, J. B., Kuster,W. C., Grand, N., Chevaillier, S., Colomb, A., Dolgorouky, C., Gros, V., Lopez, M., Sarda-Esteve, R., Holloway, J. S., Stutz, J., Petetin, H., McKeen, S., Beekmann, M., Warneke, C., Parrish, D. D., and de Gouw, J. A., 2013 Emission ratios of anthropogenic volatile organic compounds in northern mid-latitude megacities: Observations versus emission inventories in Los Angeles and Paris, *J. Geophys. Res.-Atmos.*, 118, 2041-2057.

California Air Resources Board (CARB), 2007 CEPAM: 2007 Almanac - Standard Emission Tool http://www.arb.ca.gov/app/emsinv/fcemssumcat2007.php.

California Air Resources Board (CARB), 2012 California Non-Methane Organic Gas Test Procedures http://www.arb.ca.gov/regact/2012/leviiighg2012/lev3.pdf.

California Air Resources Board (CARB), 2016 History of Air Resources Board http://www.arb.ca.gov/knowzone/history.htm.

Dallmann, T.R. and Harley, R.A., 2010. Evaluation of mobile source emission trends in the United States. *Journal of Geophysical Research: Atmospheres*, 115(D14).

EPA, 2012. Emissions Modeling Clearinghouse Speciation http://www.epa.gov/ttnchie1/emch/speciation/.

Fortin, T.J., Howard, B.J., Parrish, D.D., Goldan, P.D., Kuster, W.C., Atlas, E.L. and Harley, R.A., 2005. Temporal changes in US benzene emissions inferred from atmospheric measurements. *Environmental science and technology*, 39(6), pp. 1403-1408.

Fujita, E.M., Stockwell, W.R., Campbell, D.E., Keislar, R.E. and Lawson, D.R., 2003. Evolution of the magnitude and spatial extent of the weekend ozone effect in California's South Coast Air Basin, 1981-2000. *Journal of the Air and Waste Management Association*, 53(7), pp. 802-815.

Fujita, E. M., Campbell, D. E., Zielinska, B., Chow, J. C., Lindhjem, C. E., DenBleyker, A., ... and Lawson, D. R., 2012. Comparison of the MOVES2010a, MOBILE6. 2, and EMFAC2007 mobile source emission models with on-road traffic tunnel and remote sensing measurements. *Journal of the Air and Waste Management Association*, 62(10), 1134-1149.

Fujita, E. M., Campbell, D. E., Stockwell, W. R., and Lawson, D. R., 2013. Past and future ozone trends in California's South Coast Air Basin: Reconciliation of ambient measurements with past and projected emission inventories. *Journal of the Air and Waste Management Association*, 63(1), 54-69.

Gertler, A.W., Fujita, E.M., Pierson, W.R., Wittorff, D.N., 1996. Apportionment of NMHC Tailpipe vs Non-Tailpipe Emissions in the Fort McHenry and Tuscarora Mountain Tunnels. *Atmos. Environ.* 30, 2297-2305.

Gertler, A.W., Sagebiel, J.C., Dippel, W.A. and O'Connor, C.M., 1999. The impact of California phase 2 reformulated gasoline on real-world vehicle emissions. *Journal of the Air and Waste Management Association*, 49(11), pp. 1339-1346.

Grosjean, D., 2003. Ambient PAN and PPN in southern California from 1960 to the SCOS97-NARSTO. *Atmospheric Environment*, 37, pp. 221-238.

Haagen-Smit, A.J., 1952. Chemistry and physiology of Los Angeles smog. *Industrial and Engineering Chemistry*, 44(6), pp. 1342-1346.

Haagen-Smit, A.J. and Fox, M.M., 1954. Photochemical ozone formation with hydrocarbons and automobile exhaust. *Air Repair*, 4(3), pp. 105-136.

Harley, R.A., 2004 Chemical Composition of Vehicle-Related Volatile Organic Compound Emissions in Central California ftp://ftp.arb.ca.gov/carbis/ptsd/ccaqs_tac/00-14ccos/harleyvocspeciationdraftreport.pdf/.

Harley, R.A., Marr, L.C., Lehner, J.K. and Giddings, S.N., 2005. Changes in motor vehicle emissions on diurnal to decadal time scales and effects on atmospheric composition. *Environmental Science and Technology*, 39(14), pp. 5356-5362.

Harley, R.A., 2009 On-Road Measurement of Light-Duty Gasoline and Heavy-Duty Diesel Vehicle Emissions www.arb.ca.gov/research/apr/past/05-309.pdf.

Husar, R.B., Patterson, D.E., Blumenthal, D.L., White, W.H. and Smith, T.B., 1977. Three-dimensional distribution of air pollutants in the Los Angeles basin. *Journal of Applied Meteorology*, 16(10), pp. 1089-1096.

Koppmann, R. ed., 2008. *Volatile organic compounds in the atmosphere*. John Wiley and Sons.

Lawson, D.R., 1990. The Southern California air quality study. *Journal of the Air and Waste Management Association*, 40(2), pp. 156-165.

Los Angeles Almanac 2015 http://www.laalmanac.com/weather/we01.htm.

McDonald, B.C., Dallmann, T.R., Martin, E.W. and Harley, R.A., 2012. Long-term trends in nitrogen oxide emissions from motor vehicles at national, state, and air basin scales. *Journal of Geophysical Research: Atmospheres*, 117(D21).

Millet, D.B., Donahue, N.M., Pandis, S.N., Polidori, A., Stanier, C.O., Turpin, B.J. and Goldstein, A.H., 2005. Atmospheric volatile organic compound measurements during the Pittsburgh Air Quality Study: Results, interpretation, and quantification of primary and secondary contributions. *Journal of Geophysical Research: Atmospheres*, 110(D7).

Navazo, M., Durana, N., Alonso, L., Gómez, M.C., García, J.A., Ilardia, J.L., Gangoiti, G. and Iza, J., 2008. High Temporal Resolution Measurements of Ozone Precursors in a Rural Background Station. A Two-year Study. *Environ. Moni. Assess.* 136:53-68.

O'Connor, S., Cross, R., 2006. California's achievements in mobile source emission control. EM: Air Waste Manage. *Assoc. Mag. Environ. Manage.* 50(28). July. 28-38.

OEHHA, 2006. Atmospheric Chemistry of Gasoline-Related Emissions: Formation of Pollutants of Potential Concern. Reproductive and Cancer Hazard Assessment Branch, Office of Environmental Health Hazard Assessment, California Environmental Protection Agency. http://www.oehha.ca.gov/air/pdf/atmosChemGas.pdf.

Pang, Y., Fuentes, M. and Rieger, P., 2014. Trends in the emissions of Volatile Organic Compounds (VOCs) from light-duty gasoline vehicles tested on chassis dynamometers in Southern California. *Atmospheric Environment*, 83, 127-135.

Pang, Y., Fuentes, M. and Rieger, P., 2015. Trends in selected ambient volatile organic compound (VOC) concentrations and a comparison to mobile source emission trends in California's South Coast Air Basin. *Atmospheric Environment*, 122, pp. 686-695.

Parrish, D.D., Trainer, M., Hereid, D., Williams, E.J., Olszyna, K.J., Harley, R.A., Meagher, J.F. and Fehsenfeld, F.C., 2002. Decadal change in carbon monoxide to nitrogen oxide ratio in US vehicular emissions. *Journal of Geophysical Research: Atmospheres*, 107(D12).

Parrish, D., 2006. Critical evaluation of US on-road vehicle emission inventories. *Atmospheric Environment* 40, 2288-2300.

Parrish, D.D., Kuster, W.C., Shao, M., Yokouchi, Y., Kondo, Y., Goldan, P.D., de Gouw, J.A., Koike, M., Shirai, T., 2009. Comparison of air pollutant emissions among mega-cities. *Atmospheric Environment*, 43, 6435-6441.

Parrish, D.D., Singh, H.B., Molina, L. and Madronich, S., 2011. Air quality progress in North American megacities: A review. *Atmospheric Environment*, 45(39), pp. 7015-7025.

Petrea, M., 2007. Emissions of non-methane volatile organic compounds (NMVOC) from vehicular traffic in Europe, Dissertation of the Bergische Universität, Wuppertal.

Piccot, S.D., Watson, J.J. and Jones, J.W., 1992. A global inventory of volatile organic compound emissions from anthropogenic sources. *Journal of Geophysical Research: Atmospheres*, 97(D9), pp. 9897-9912.

Pierson, W.R., Schorran, D.E., Fujita, E.M., Sagebiel, J.C., Lawson, D.R. and Tanner, R.L., 1999. Assessment of nontailpipe hydrocarbon emissions from motor vehicles. *Journal of the Air and Waste Management Association*, 49(5), pp. 498-519.

Pollack, I.B., Ryerson, T.B., Trainer, M., Neuman, J.A., Roberts, J.M. and Parrish, D.D., 2013. Trends in ozone, its precursors, and related secondary oxidation products in Los Angeles, California: A synthesis of measurements from 1960 to 2010. *Journal of Geophysical Research: Atmospheres*, 118(11), pp. 5893-5911.

Propper, R., Wong, P., Bui, S., Austin, J., Vance, W., Alvarado, Á., Croes, B. and Luo, D., 2015. Ambient and Emission Trends of Toxic Air

Contaminants in California. *Environmental science and technology*, 49 (19), pp. 11329-11339.

Renzetti, N.A., 1956. Ozone in the Los Angeles atmosphere. *The Journal of Chemical Physics*, 24(4), pp. 909-909.

Rubin, J.I., Kean, A.J., Harley, R.A., Millet, D.B., Goldstein, A.H., 2006. Temperature dependence of volatile organic compound evaporative emissions from motor vehicles. *J. Geophys. Res.* 111, D03305.

Stemmler, K., Bugmann, S., Buchmann, B., Reimann, S., Staehelin, J., 2005. Large decrease of VOC emissions of Switzerland's car fleet during the past decade: results from a highway tunnel study. *Atmospheric Environment* 39, 1009-1018.

Warneke, C., J. A. de Gouw, J. S. S. Holloway, J. Peischl, T. B. B. Ryerson, E. L. Atlas, D. R. Blake, M. K. Trainer, and D. D. Parrish, 2012. Multi-Year Trends in Volatile Organic Compounds in Los Angeles, California: Five Decades of Decreasing Emissions, *J. Geophys. Res.*, doi:10.1029/2012JD0 17899.

Watson, J. G.; Chow, J. C.; Fujita, E. M. 2001. Review of volatile organic compound source apportionment by chemical mass balance. *Atmos. Environ.* 35, 1567-1584.

In: Volatile Organic Compounds
Editor: Julian Patrick Moore
ISBN: 978-1-63485-370-5

Chapter 5

OXIDATION OF VOC OVER CRYPTOMELANE CATALYSTS: DOPING WITH GOLD AND ALKALI

S. A. C. Carabineiro[*], V. P. Santos, O. S. G. P. Soares, M. F. R. Pereira, J. J. M. Órfão and J. L. Figueiredo

Laboratory of Separation and Reaction Engineering –
Laboratory of Catalysis and Materials (LSRE-LCM),
Faculty of Engineering, University of Porto, Porto, Portugal

ABSTRACT

Volatile organic compounds (VOC) are hazardous environmental pollutants originated from different sources, such as petroleum refineries, fuel storage, motor vehicles, painting and printing activities. VOCs emissions are regulated due to their potential damages to human health and the environment.

Catalytic oxidation to CO_2 and H_2O is an environmentally friendly technology for VOC abatement that needs low temperatures (around 250-500ºC) and causes less NOx formation, compared to conventional thermal oxidation, which operates at higher temperatures (650-1100ºC). Several catalysts have been used for this purpose.

In this work, cryptomelane-type manganese oxides were tested for the oxidation of ethyl acetate and toluene, two common VOCs. Catalysts were synthesized by redox reaction under acid and reflux conditions. Different metals (cerium, cesium and lithium) were incorporated into the

[*] Corresponding Author. Email: scarabin@fe.up.pt.

tunnel structure of cryptomelane by the ion-exchange technique. Gold was loaded onto these materials (1% wt.) by a double impregnation method. The obtained catalysts were characterized by X-ray diffraction, high-resolution transmission electron microscopy, energy-dispersive X-ray diffraction and temperature programmed reduction.

It was found that addition of Cs and Li to cryptomelane was beneficial for ethyl acetate oxidation, but addition of Ce is detrimental, which is related to the reducibility of the materials. Addition of gold does not improve the catalytic activity, and in case of Li and Cs modified samples, it has even an unfavourable effect. This effect can be related to the large gold particle size found for these samples, well known to be inversely related with catalytic activity. Addition of Li to cryptomelane was beneficial for toluene oxidation, but adding Cs and Ce was disadvantageous. Loading with gold also did not show considerable improvement, which may also be related to particle size effects. The activity for both VOCs abatement was correlated with sample reducibility.

Keywords: gold, cryptomelane, alkali-doping, ethyl acetate, toluene

INTRODUCTION

Volatile organic compounds (VOCs) are hazardous environmental pollutants originated from stationary and mobile sources, such as painting, printing, petroleum refineries, fuel storage and motor vehicles [1, 2]. Examples of common VOCs are acetone, ethanol, propanol, ethyl acetate, buthyl acetate, benzene, toluene and methyl chloride [3]. Halogenated organic compounds, such as chlorofluorocarbons (CFCs) and hydrochlorofluorocarbons (HCFCs), are other examples of VOCs. Some of these compounds are malodorous or carcinogenic. When emitted into the atmosphere, VOCs are capable of undergoing reactions with NO_x, in the presence of sunlight, to form photochemical oxidants, such as ozone (O_3), peroxyacetyl nitrate (PAN) and hydrogen peroxide (H_2O_2). For this reason, VOCs are considered as precursors for the potential formation of tropospheric ozone, which at ground level can damage human health, fauna, flora and buildings. In addition, VOCs are responsible for the formation of secondary particulates in the atmosphere that lead to smog formation. Micro or even nanoparticulates are formed by nucleation processes together with NO_x, SO_x and NH_3, which can penetrate into human lungs, causing premature deaths and aggravating respiratory conditions such as asthma [4]. Therefore, in order to minimize the negative impacts, VOCs are worldwide regulated in many sectors of industry and traffic.

Several VOC control technologies have been developed. Catalytic oxidation seems one of the most promising. It is an environmentally friendly technology that requires low temperatures (around 250-500ºC) and causes less NO_x formation, compared to conventional thermal oxidation that requires high operation temperatures (650-1100ºC) [2, 3, 5].

Metal oxides can be used as heterogeneous catalysts for VOC oxidation. Commonly used metal oxides include manganese dioxide [6-22], copper oxide [23-26], nickel oxide [23, 25-27] and cerium oxide [16, 28, 29]. It has been reported that metal oxide based catalysts are more resistant to poisoning phenomena, but are generally less active than supported precious metals in oxidizing VOC streams [30, 31]. However, mixed oxides show good activities in catalytic oxidation of VOCs [28, 29, 32-39].

In particular, octahedral molecular sieves (OMS) have received considerable attention, due to their amazing catalytic properties [12, 40-43]. These materials contain open frameworks that consist in edge and corner shared MnO_6 octahedra, and their internal pores are occupied by cations and water molecules. Cryptomelane-type manganese oxides (K-OMS-2) have 2×2 arrays of MnO_6 octahedra tunnel structures [12, 41]. Potassium cations and small amounts of water inside the tunnels can stabilize the structure, and be partially ion-exchanged with other cations with appropriate sizes. The average manganese oxidation state is around 3.8, which corresponds to a higher amount of Mn (IV) and less Mn (III) [12, 40-44]. These features make cryptomelane materials very attractive for oxidation reactions.

The electronic and catalytic properties of cryptomelane can be further improved by adding other metal cations inside the tunnels or into the framework by ion-exchange [12, 45-47] or by substitution during synthesis [48, 49]. The introduction of alkali metal into the tunnel of cryptomelane can significantly modify the surface acid-base properties [10, 50]. There are reports of OMS materials modified with Cu [45, 51-53], Co [45, 54], Ce [21], Ni [45], Ag [55-58], Pd [54, 59], Au [60], among others. These materials have been used as catalysts for the total oxidation of VOCs [10, 12, 13, 15, 17, 19, 20, 61-74]. The high activity of OMS materials has been attributed to the presence of the redox couple Mn(III)/Mn(IV), high mobility of lattice oxygen, open structure and high hydrophobicity [12].

In previous works, we used cryptomelane-type catalysts for the oxidation of VOCs (ethyl acetate, toluene and ethanol) [11-13, 18, 19] and studied the effect of adding different amounts of Cs and Li to cryptomelane on the oxidation of ethyl acetate [10, 20]. The effect of gold addition to a Ce doped cryptomelane was also reported for the oxidation of CO [21], and the effect of Au on several manganese oxides for CO [75] and VOCs [16] oxidation. In this work, cryptomelane modified with Ce, Cs and Li, alone and loaded with Au is tested for VOC (ethyl acetate and toluene) oxidation.

EXPERIMENTAL

Catalyst Preparation

Cryptomelane-type manganese oxide (K-OMS-2) was synthesized using a reflux approach in acidic medium developed by Luo et al. [61]. Cerium, cesium and lithium were incorporated into the tunnel structure by ion-exchange (samples Ce-K-OMS-2, Cs-K-OMS-2 and Li-K-OMS-2, respectively): 1 g of cryptomelane was stirred with 30 mL of 0.1 M solution of the respective nitrate precursors (Aldrich) during 7 days. The solid obtained was filtered and washed with distilled water, followed by drying at 100ºC and calcination at 450ºC in air for 4.5 h.

Gold was loaded onto the supports by a double impregnation method [16, 75-77], which briefly consists in impregnation with a solution of $HAuCl_4$ (using sonication), followed by addition of an aqueous solution of Na_2CO_3 (1 M), under constant ultrasonic stirring. The resulting solid was washed repeatedly with distilled water for chloride removal (which is well known to cause sinterisation of Au nanoparticles, thus turning them inactive [78-81]), and dried in the oven at ~110ºC overnight.

Characterisation Techniques

Supports were characterized by N_2 adsorption at −196ºC and temperature programmed reduction (TPR). Selected samples were also analysed by scanning electron microscopy (SEM) and X-ray diffraction (XRD). Au samples were characterized by SEM, TPR and by transmission electron microscopy (TEM). Chemical composition was determined with energy-dispersive X-ray spectroscopy (EDS).

BET surface areas were calculated from the N_2 adsorption isotherms at -196ºC obtained in a Quantachrome Instruments Nova 4200e. All samples were previously degassed before analysis at 160ºC, for 5 h.

TPR experiments were carried out in a fully automated AMI-200 Catalyst Characterization Instrument (Altamira Instruments) under H_2 atmosphere, to acquire information on the reducibility of the samples.

Surface analysis for morphological characterization was carried out by SEM, using a FEI Quanta 400 FEG ESEM (15 keV) electron microscope. The sample powders were mounted on a double-sided adhesive tape and observed at different magnifications under two different detection modes: secondary and back-scattered electrons. EDS confirmed the nature of the components.

The samples composition and average oxidation state (AOS) of manganese on the surface were determined by X-ray photoelectron spectroscopy (XPS) using a Kratos AXIS Ultra HSA, with VISION software for data acquisition and CASAXPS software for data analysis. The effect of the electric charge was corrected by the reference of the carbon peak (284.6 eV).

XRD analysis was carried out in a PAN'alytical X'Pert MPD equipped with an X'Celerator detector and secondary monochromator (Cu Kα = 0.154 nm, 50 kV, 40 mA). The collected spectra were analysed by Rietveld refinement using the PowderCell software, allowing the determination of the crystallite sizes.

The structural characterization of the samples was performed using a JEOL 2010F Field Emission Gun Transmission Electron Microscope (FEG-TEM), working in both High Resolution TEM (HR-TEM) and Scanning TEM (STEM) modes at 200 kV accelerating voltage.

Catalytic Tests

The catalytic reactions were performed in a U-shaped quartz tube fixed-bed reactor with 6 mm internal diameter, placed inside a temperature controlled electrical furnace, with a total air flow rate of 500 cm^3/min (measured at room temperature and atmospheric pressure), corresponding to a space velocity of 60,000 h^{-1} (determined in terms of total bed volume), having a composition of 1000 mg_{Carbon}/m^3 (~466.7 ppmV of ethyl acetate or ~226 ppmV of toluene). 50 mg of catalyst (with particle sizes between 0.2 and 0.5 mm) was used in each experiment. The catalyst was mixed thoroughly with an inert (SiC, carborundum) with particle sizes between 0.2 and 0.5 mm. The total volume of the mixture of catalyst sample and inert was about 0.5 cm^3. A pre-treatment in air was carried out before the catalytic reaction by heating from room temperature up to 350ºC at 10ºC/min. Two cycles of increasing (at a heating rate of 2ºC/min) and decreasing temperature were performed for each catalyst. The extent of VOC oxidation was evaluated by continuously monitoring CO_2 formation with a non-dispersive infrared (NDIR) sensor (Vaisala GMP222). The concentration of VOC in the effluent was also measured with a total VOC analyser MiniRAE2000. The catalytic performance is presented as conversion into CO_2, X_{CO2}, obtained by the following equation:

$$X_{CO_2} = \frac{F_{CO_2}}{\nu \cdot F_{in,\,VOC}},$$

where $F_{in,\,VOC}$ is the inlet molar flow rate of VOC, F_{CO2} is the outlet molar flow rate of CO_2, and ν is the number of carbon atoms in the VOC molecule (for ethyl acetate, $\nu = 4$; for toluene, $\nu = 7$).

RESULTS AND DISCUSSION

Textural Characterization of Supports

K-OMS-2 is characterized by a BET surface area of 55 m^2/g. The incorporation of dopants (Ce, Cs and Li, Au) did not change significantly the textural properties of cryptomelane.

XRD

The crystalline structure of the materials synthesized in this work was characterizaed by XRD, (Figure 1). All materials are monophasic and showed the presence of the cryptomelane phase (space group I4/m; JCPDS 42-1348), except the Ce-OMS-2, which also contained CeO_2 (cerianite), as seen in Figure 1a. Gold loaded materials were not analysed by XRD as usually loadings of ~1% Au wt. are not detected by XRD [82-84]. Incorporation of Cs or Li into cryptomelane shifts the diffraction peaks (310 and 211) to lower angles, as can be seen in Figures 1b and c), while inclusion of Ce has the opposite effect. Shifts are expected when the dopants are included in the lattice of cryptomelane structure, promoting expansion and/or contractions of the lattice [85].

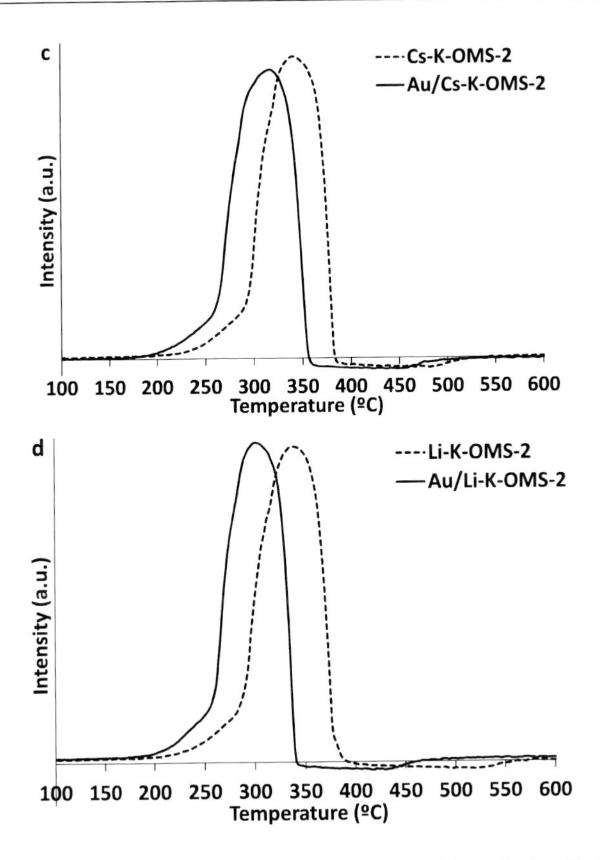

Figure 2. H_2-TPR patterns for cryptomelane alone (a) and doped with Ce (b), Cs (c) and Li (d), with and without Au.

In a previous work, upon incorporation of with a solution of 0.5 M Ce, two overlapping peaks centered at ~350ºC and ~420ºC, and a low intensity peak at high temperatures (500ºC) were observed [21]. Interestingly, in this work, with a solution of 0.1 M Ce, only a broad peak with a shoulder at low temperatures is observed (Figure 2b), similarly to what is found for Cs-cryptomelane (Figure 2c) and Li- cryptomelane (Figure 2d) and cryptomelane (Figure 2a). However, the peak of Ce-K-OMS2 is shifted to higher temperatures (compared to that of K-OMS-2, Figure 2a), indicating that reduction is hindered by the presence of Ce, unlike what occurs with Cs and Li, as their incorporation into cryptomelane facilitates reduction. Addition of Au to all materials causes a shift to lower temperatures (Figure 2) indicating that reduction is favoured when Au is present. This is not so evident for OMS-2, as addition of Au caused only a small shift to lower temperatures (Figure

2a), but is more important for Ce-K-OMS-2 (Figure 2b), and particularly for samples loaded with Cs (Figure 2c) and Li (Figure 2d), where the effect of Au is more pronounced.

Figure 3. SEM images of pure cryptomelane (a) and doped with Ce (b), Cs (c) and Li (d).

SEM and TEM

The materials were also analysed by SEM, and they all have a similar and homogenous morphology, consisting of long nanofibers that resemble “needles” of “straw” (Figure 3). Au-loaded samples were analysed by HRTEM (Figure 4). Cryptomelane with Au is shown in Figures 4a and 4b. Gold nanoparticles from 8 to 18 nm can be observed on the cryptomelane fibres. Gold nanoparticles are deposited on the outer surface of the cryptomelane, and not in the OMS-2 channels, as shown before with Ag [55] and with Au [60]. Large sizes were reported in literature for gold nanoparticles on cryptomelane-type catalysts using other deposition methods [21, 60].

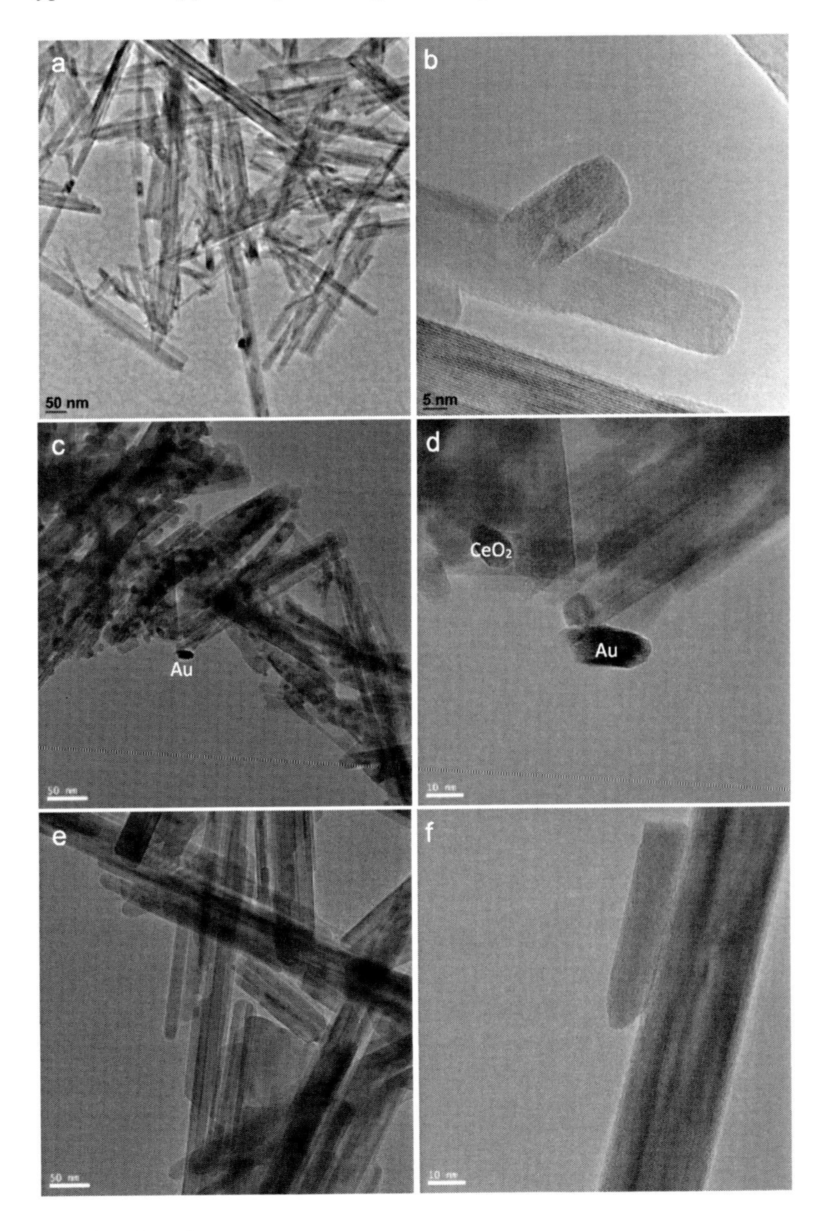

Figure 4. (Continued).

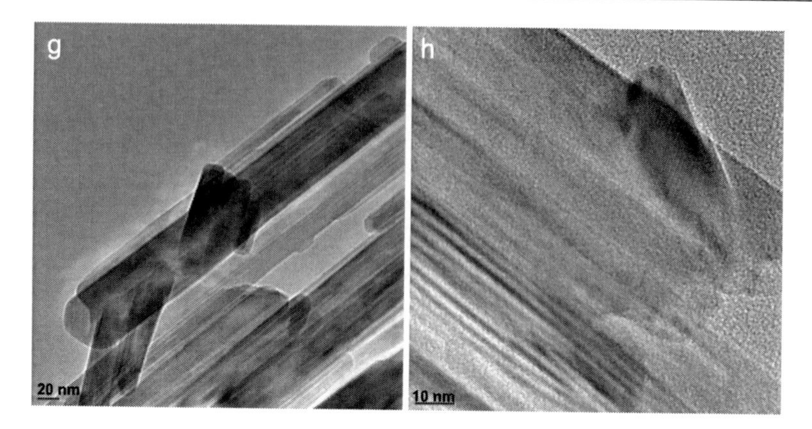

Figure 4. HR-TEM images of gold loaded materials: cryptomelane (a, b) and cryptomelane doped with Ce (c, d), Cs (e, f) and Li (g, h). Images on the right (b, d, e, f) are closer details.

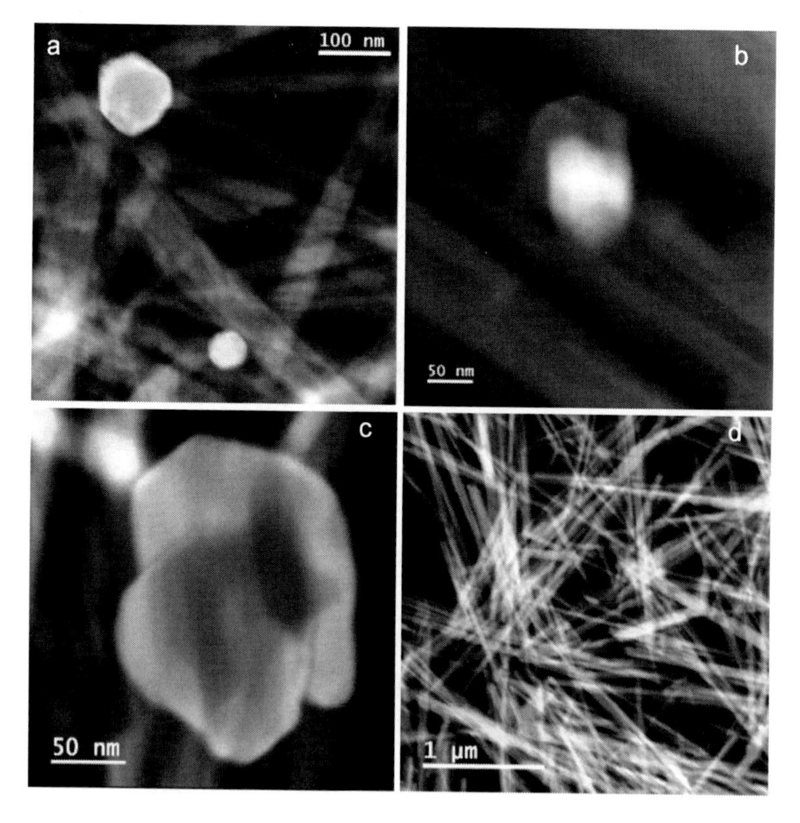

Figure 5. STEM images of pure cryptomelane (a) and doped with Ce (b), Cs (c) and Li (d).

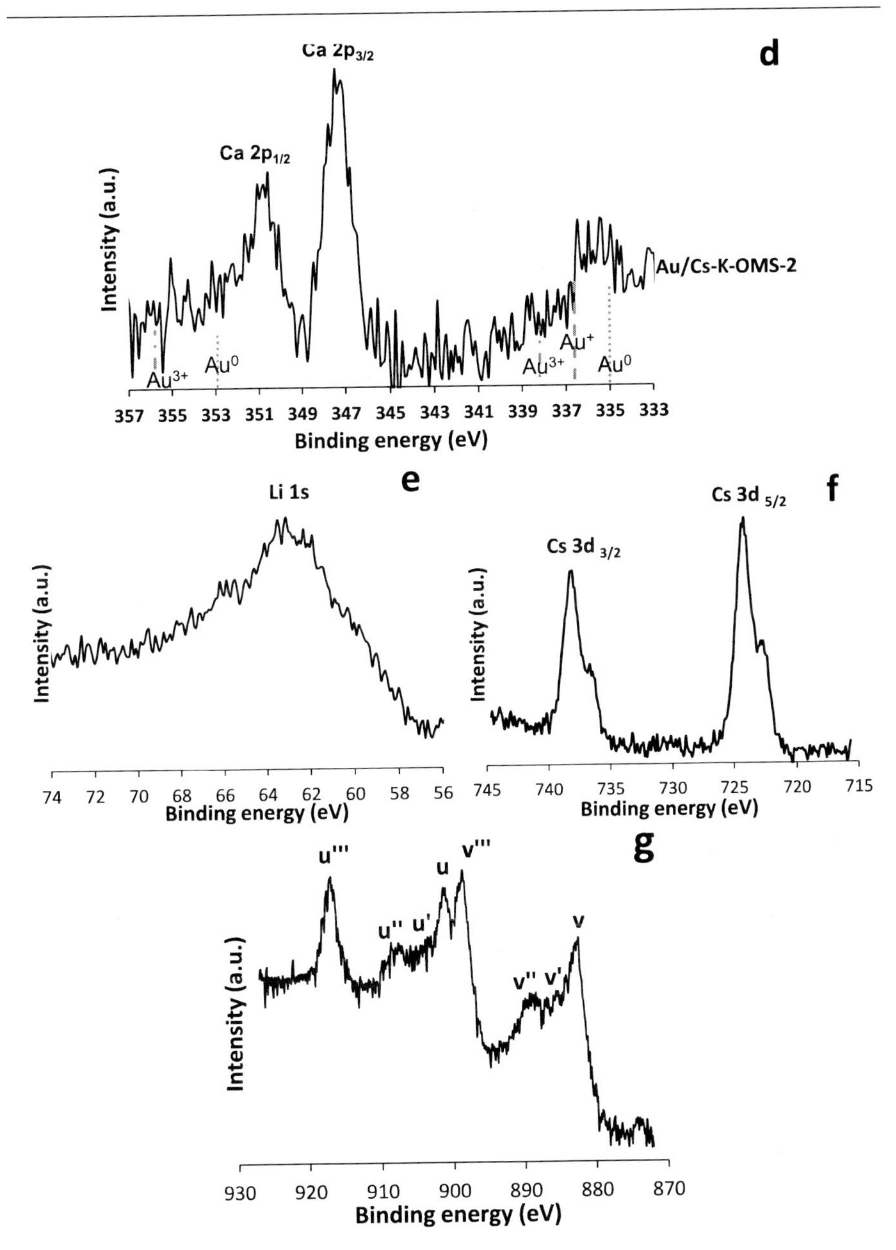

Figure 7. XPS Mn 2p (a) Mn 3p (b) cryptomelanes doped with Ce, Cs and Li. XPS Au 4f (c) and superimposium of Au 4d and Mn 3s (d) of the same materials loaded with Au. XPS Li 1s (e), Cs 3d (f), and Ce 3d (g) of the Au/cryptomelanes doped with Li, Cs and Ce, respectively.

XPS

The oxidation state of Mn on cryptomelane samples was assessed by analysing the XPS Mn2p, Mn3s and Mn3p spectra. The spectra of Mn and other transition metals often shown the presence of satellite peaks (shake up and shake down processes) and multiplet splitting, resulting from the spin and spatial angular momentum coupling of the open core and the open d shell of the transition metal (exchange splitting). These couplings originate final-state multiplets of different binding energies, and can be explained in terms of configuration – interaction calculation including intra-shell electron correlation, charge transfer and final state screening effects. Multiplet structures can also be influenced by covalence and ligand coordination and ligand electronegativity [20, 90-92].

The Mn2p spectra are shown in Figure 7a. The values of binding energy are characteristic of the values reported in the literature for cryptomelane [93]. The position of these peaks does not change significantly when Mn changes oxidation state, but the satellite peak of Mn2p is sensitive to Mn oxidation state [94, 95]). As shown in Figure 7a, only one very low intensity peak can be found at ~665 eV as the others are overlapped by $Mn2p_{1/2}$ peak, causing some asymmetry [90, 92]. Therefore, we can conclude that there is no Mn(II) present. Loading the materials with Au does not change much the peak appearance (not shown).

Figure 7b shows XPS Mn3p for criptomelane materials and Figure 7c shows Mn3s for materials loaded with gold (which superimposes with Au 4f). Results are typical for species with a net parallel spin in the valence state interacting with the final state. As explained in a previous work [20], the Mn valence band configuration is $3d^5 4s^2$. Photoelectron transitions are allowed between the initial state $3s^2 3p^6 3d^N$ (where N = 3, 4 and 5 for Mn(IV), Mn(III), and Mn(II), respectively) and the final state $3s^1 3p^6 3d^{N+1}$. The spin vector of the 3s remaining electron can be parallel or anti-parallel to the 3d electrons spin. When spins are parallel, electrons of 3s and 3d orbitals exchange and the binding energy is lowered. Therefore, lower energy peak is due to parallel spins, while the higher energies peaks are from the anti-parallel spins. The Mn 3s multiplet splitting can be used to calculate the average oxidation state (AOS) of manganese, according to the relationship [92]: $AOS = 8.95 - 1.13\,\Delta E$ (eV), where ΔE is the energy difference between the main peak and its satellite (multiplet splitting). The results for AOS are 3.7 (for pure cryptomelane), 3.8 (Li-cryptomelane) and 3.9 (Ce and Cs doped cryptomelane). Those values are in agreement with the expected mixed valence state of Mn in cryptomelane materials. The Mn 3p peak can also indicate the oxidation state of Mn. If it shifts to higher energies, then a higher oxidation state is present. The main Mn 3p peak can be found at ~50 eV, showing that Mn(III) and Mn(IV) are there [95], confirming what was found with the Mn 2p and Mn 3s results. Spectra are very similar for pure cryptomelane and samples doped with Ce, Cs and Li and only small shifts can be found which are responsible for the small increase of AOS. Not many differences were found for samples loaded with Au (not shown).

In order to determine the oxidation state of gold, the Au 4f spectra were measured for the Au-containing catalysts, however superimposition of the Mn 3s peak was found (Figure 7c). The Au 4d line is less intense than Au 4f and Ca 2p peak was also caught in the measurements (Figure 7d). That is most likely due to the fact that these doped samples need a long preparation time, that is, 7 day stirring in a glass flask (see Experimental), and Ca comes from the glass. Only a small peak is observed between 336 and 334 eV that can be attributed to Au, but is it not possible to determine if it is in the Au^+ or Au^0 state. We tried to measure Au 4p, but as this line is even less intense than Au 4d, only noise was detected (not shown), as also happened in a previous work [26].

Figures 7e, f, g show the XPS Li 1s, Cs 3d, and Ce spectra for cryptomelanes doped with Li, Cs, and Ce, respectively. The peak binding energies are according to what is expected from literature [77, 96, 97]. Li 1s spectrum (Figure 7e) is greatly influenced by the proximity of Mn 3p (not shown), that is a very intense and complex peak. In terms of Ce spectrum, it can be resolved into eight components with the assignment of each component defined in the figure (v represent the Ce $3d_{5/2}$ contributions and u represent the Ce $3d_{3/2}$ contributions) [28, 77, 98]. For $3d_{5/2}$ of Ce^{4+}, a mixing of the Ce $3d^94f^2L^{n-2}$ and Ce $3d^94f^1L^{n-1}$ states produces the peaks labeled v and v',' and the Ce $3d^94f^0L^n$ final state forms the peak v''.' For $3d_{5/2}$ of Ce^{3+}, the Ce $3d^94f^2L^{n-1}$ and Ce $3d^94f^1L^n$ states correspond to peaks v and v.' For Ce $3d_{3/2}$ level with the u structure, the same assignment can be carried out. A relationship between the area of CeO_2 in the Ce 3d spectrum and the % of u''' peak (which arises from a transition of the 4f" final state from the 4f" initial state) was established by Shyu et al. [98], from which the % CeO_2 can be calculated using the equation: % CeO_2 = % u''' / 0.144. It was reported earlier that upon gold loading, the amount of surface oxygen is increased on ceria/cryptomelane [21]. Similar results were obtained now, as the amount of Ce^{3+} increased from ~10% to ~35% on the catalyst, upon gold loading.

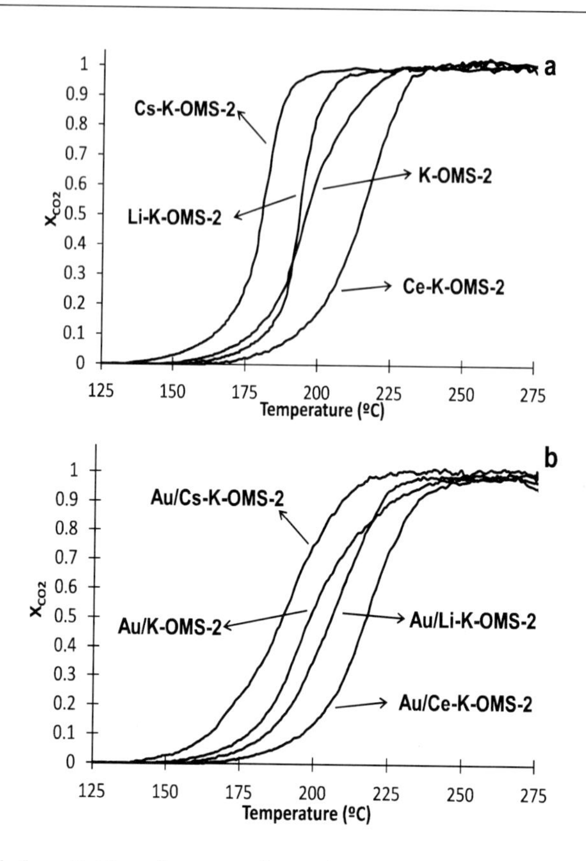

Figure 8. Catalytic activities of cryptomelanes doped with Ce, Cs and Li (a) and the same materials loaded with Au (b) towards total oxidation of ethyl acetate. Curves for Cs and Li doped cryptomelanes in a) were taken from [20].

Catalytic Results

The catalytic activities for the total oxidation of ethyl acetate (Figure 8) and toluene (Figure 9) of cryptomelanes doped with Ce, Cs and Li (a) and the same materials loaded with Au (b) were evaluated in order to determine the effect of the different dopants. In a previous work, the positive effect of adding different amounts of Cs and Li to cryptomelane on the oxidation of ethyl acetate was reported [10, 20]. It was found that the activity increase was explained by the redox and basic properties of the modified materials. Those results are now compared with Ce loaded cryptomelane (Figure 8a) and it can be seen that Ce has the opposite effect, as the activity decreases. Figure 8b shows the same materials loaded with Au. It can be seen that adding gold does not improve the catalytic activity, and in case of Li and Cs, it has even a detrimental effect. That is most likely due to the large gold particle size found for these samples (Figures 4-6), well known to be inversely related to catalytic activity. The same effect was observed in previous works with manganese oxides for CO [75] and VOCs [16] oxidation, as gold nanoparticles were also of large size.

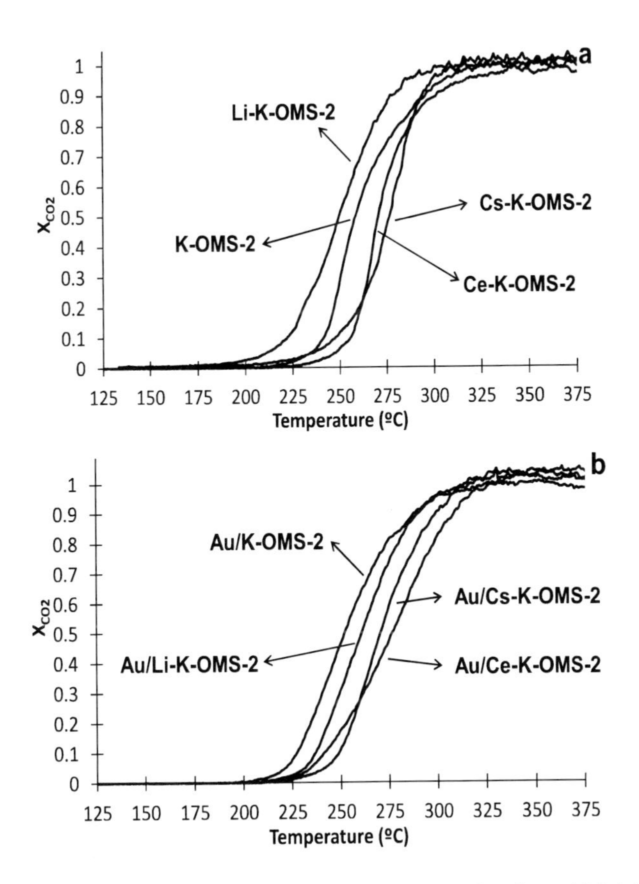

Figure 9. Catalytic activities of cryptomelanes doped with Ce, Cs and Li (a) and the same materials loaded with Au (b) towards total oxidation of toluene.

Figure 9a shows the catalytic activities for toluene oxidation for the doped cryptomelane samples. Adding Li shows an improvement, as also observed with ethyl acetate (Figure 8a), but adding Cs and Ce is detrimental. Loading with Au does not show much improvement (Figure 9b), when compared with the unloaded materials (Figure 9a). The curves for the Cs doped material are very similar. There is a slight improvement for the Ce-cryptomelane and cryptomelane samples, especially at lower temperatures. That can be due to the increase in the amount of oxygen, as found by XPS. A positive effect on adding Au to Ce-cryptomelane was reported for CO oxidation in a previous work [21]. It was found that cerium increases the number of defect sites that anchor gold nanoparticles, enhances the charge transfer between Au and cryptomelane and promotes the stabilization of gold atoms and clusters [21]. The activity of Au/CeO_2 catalyst has been explained by the ability of gold nanoparticles to increase the mobility of lattice oxygen by weakening the Ce–O bonds adjacent to Au atoms, which increases the lattice oxygen donating ability of CeO_2 to oxidize the VOC molecule through a Mars-van Krevelen reaction mechanism [99]. However, Figure 9b also shows that there is a detrimental effect for Li containing sample, as also found with ethyl acetate (Figure 8b). Once again that is most likely due to the fact that this sample is the one that showed the largest gold particle sizes (Figure 6).

Interestingly, the observed trend in catalytic activity for gold loaded samples (Figures 8b and 9b) can be related to the reducibility of catalysts (Figure 2), showing that the redox behavior of the materials influences their VOC oxidation activity. It can be seen that the less active sample (Ce-cryptomelane, with or without Au) is also the less reductible. The impact of redox properties on the oxidation performance of metal oxides has been reported for several systems in the past (e.g., [16, 18, 25, 26, 28, 29, 39, 77]).

CONCLUSION

Cryptomelane-type manganese oxides loaded with Ce, Cs and Li were tested for the oxidation of ethyl acetate and toluene. The incorporation of Cs and Li onto cryptomelane was beneficial for ethyl acetate oxidation, but addition of Ce is detrimental. Addition of Li to cryptomelane was beneficial for toluene oxidation, but adding Cs and Ce was unfavourable.

Addition of Au did not improve the catalytic activity, and in some cases it had even a negative effect. That is most likely due to the large gold particle size found for these samples, well known to be inversely related to catalytic activity. The activity for both VOCs abatement was related with sample reducibility.

ACKNOWLEDGMENTS

This work was supported by Fundação para a Ciência e a Tecnologia (FCT) and FEDER (Ph.D. grant SFRH/BD/23731/2005 for VPS). SACC is grateful to FCT for CIENCIA 2007 and Investigador FCT (IF/01381/2013/CP1160/CT0007) grants, with financing from the European Social Fund and the Human Potential Operational Program. This work was co-financed by project POCI-01-0145-FEDER-006984 – Associate Laboratory LSRE-LCM funded by FEDER funds through COMPETE2020 - Programa Operacional Competitividade e Internacionalização (POCI) – and by national funds through FCT. The authors acknowledge Dr Carlos Sá (CEMUP) for assistance in SEM and XPS analyses and Dr. Alessandro Benedetti (University of Vigo) for assistance with the TEM measurements.

REFERENCES

[1] Tucker WG. Volatile Organic Compounds. 2004. In: Indoor Air Quality Handbook [Internet]. New York: Digital Engineering Library @ McGraw-Hill.

[2] Huang H, Xu Y, Feng Q, Leung DYC. Low temperature catalytic oxidation of volatile organic compounds: a review. *Catalysis Science & Technology*. 2015;5(5):2649-69.

[3] Khan FI, Ghoshal AK. Removal of Volatile Organic Compounds from polluted air. *Journal of Loss Prevention in the Process Industries.* 2000;13:527-45.

[4] The Clean Air for Europe (CAFE) Programme: Towards a Thematic Strategy for Air Quality. COM(2001) 245 final ed. Brussels: Comission of the European Communities; 2001.

[5] Moretti EC. Reduce VOC and HAP Emissions. *CEP magazine*. 2002 June 2002:30-40.

[6] Cellier C, Ruaux V, Lahousse C, Grange P, Gaigneaux EM. Extent of the participation of lattice oxygen from [gamma]-MnO2 in VOCs total oxidation: Influence of the VOCs nature. *Catal Today*. 2006;117(1-3):350-5.

[7] Peluso MA, Gambaro LA, Pronsato E, Gazzoli D, Thomas HJ, Sambeth JE. Synthesis and catalytic activity of manganese dioxide (type OMS-2) for the abatement of oxygenated VOCs. *Catal Today*. 2008;133:487-92.

[8] Sinha AK, Suzuki K, Takahara M, Azuma H, Nonaka T, Suzuki N, et al. Preparation and Characterization of Mesostructured gamma-Manganese Oxide and Its Application to VOCs Elimination. *J Phys Chem C.* 2008;112(41):16028-35.

[9] Raciulete M, Afanasiev P. Manganese-containing VOC oxidation catalysts prepared in molten salts. *Appl Catal A: Gen.* 2009;368(1-2):79-86.

[10] Santos VP, Pereira MFR, Órfão JJM, Figueiredo JL. Catalytic oxidation of ethyl acetate over a cesium modified cryptomelane catalyst. *Appl Catal B: Environ.* 2009;88(3-4):550-6.

[11] Santos VP, Pereira MFR, Órfão JJM, Figueiredo JL. Synthesis and Characterization of Manganese Oxide Catalysts for the Total Oxidation of Ethyl Acetate. *Top Catal.* 2009;52(5):470-81.

[12] Santos VP. Catalytic Oxidation of Volatile Organic Compounds, Ph.D. Thesis [Ph.D. Thesis]. Porto: Porto University; 2010.

[13] Santos VP, Bastos SST, Pereira MFR, Órfão JJM, Figueiredo JL. Stability of a cryptomelane catalyst in the oxidation of toluene. *Catal Today.* 2010;154(3-4):308-11.

[14] Pérez H, Navarro P, Delgado JJ, Montes M. Mn-SBA15 catalysts prepared by impregnation: Influence of the manganese precursor. *Appl Catal A: Gen.* 2011;400(1–2):238-48.

[15] Sanz O, Delgado JJ, Navarro P, Arzamendi G, Gandia LM, Montes M. VOCs combustion catalysed by platinum supported on manganese octahedral molecular sieves. *Appl Catal B: Environ.* 2011;110:231-7.

[16] Bastos SST, Carabineiro SAC, Órfão JJM, Pereira MFR, Delgado JJ, Figueiredo JL. Total oxidation of ethyl acetate, ethanol and toluene catalyzed by exotemplated manganese and cerium oxides loaded with gold. *Catal Today.* 2012;180(1):148-54.

[17] Genuino HC, Dharmarathna S, Njagi EC, Mei MC, Suib SL. Gas-Phase Total Oxidation of Benzene, Toluene, Ethylbenzene, and Xylenes Using Shape-Selective Manganese Oxide and Copper Manganese Oxide Catalysts. *J Phys Chem* C. 2012;116(22):12066-78.

[18] Santos VP, Pereira MFR, Órfão JJM, Figueiredo JL. The role of lattice oxygen on the activity of manganese oxides towards the oxidation of volatile organic compounds. *Appl Catal B: Environ.* 2010;99(1-2):353-63.

[19] Santos VP, Pereira MFR, Órfão JJM, Figueiredo JL. Mixture effects during the oxidation of toluene, ethyl acetate and ethanol over a

cryptomelane catalyst. *Journal of Hazardous Materials*. 2011;185(2-3):1236-40.

[20] Santos VP, Soares OSGP, Bakker JJW, Pereira MFR, Órfão JJM, Gascon J, et al. Structural and chemical disorder of cryptomelane promoted by alkali doping: Influence on catalytic properties. *J Catal.* 2012;293:165-74.

[21] Santos VP, Carabineiro SAC, Bakker JJW, Soares OSGP, Chen X, Pereira MFR, et al. Stabilized gold on cerium-modified cryptomelane: Highly active in low-temperature CO oxidation. *J Catal.* 2014;309(0):58-65.

[22] Yodsa-nga A, Millanar JM, Neramittagapong A, Khemthong P, Wantala K. Effect of manganese oxidative species in as-synthesized K-OMS 2 on the oxidation of benzene. *Surface & Coatings Technology*. 2015;271:217-24.

[23] Pooperasupong S, Caussat B, Damronglerd S. Air Pollution Control by Oxidation of Aromatic Hydrocarbon over Supported Metal Oxide. *ScienceAsia*. 2008;34(1):119.

[24] Huang Y-C, Luo C-H, Yang S, Lin Y-C, Chuang C-Y. Improved Removal of Indoor Volatile Organic Compounds by Activated Carbon Fiber Filters Calcined with Copper Oxide Catalyst. *CLEAN - Soil, Air, Water*. 2010;38(11):993-7.

[25] Chen X, Carabineiro SAC, Bastos SST, Tavares PB, Órfão JJM, Pereira MFR, et al. Exotemplated copper, cobalt, iron, lanthanum and nickel oxides for catalytic oxidation of ethyl acetate. *Journal of Environmental Chemical Engineering*. 2013;1(4):795-804.

[26] Carabineiro SAC, Chen X, Martynyuk O, Bogdanchikova N, Avalos-Borja M, Pestryakov A, et al. Gold supported on metal oxides for volatile organic compounds total oxidation. *Catal Today*. 2015;244:103-14.

[27] Solsona B, Garcia T, Aylón E, Dejoz AM, Vázquez I, Agouram S, et al. Promoting the activity and selectivity of high surface area Ni–Ce–O mixed oxides by gold deposition for VOC catalytic combustion. *Chemical Engineering Journal*. 2011;175:271-8.

[28] Konsolakis M, Carabineiro SAC, Tavares PB, Figueiredo JL. Redox properties and VOC oxidation activity of Cu catalysts supported on Ce1-xSmxOδ mixed oxides. *Journal of Hazardous Materials*. 2013;261:512-21.

[29] Carabineiro SAC, Chen X, Konsolakis M, Psarras AC, Tavares PB, Órfão JJM, et al. Catalytic oxidation of toluene on Ce-Co and La-Co

mixed oxides synthesized by exotemplating and evaporation methods. *Catal Today*. 2015;244:161-71.

[30] Heneghan CS, Hutchings GJ, Taylor SH. The destruction of volatile organic compounds by heterogeneous catalytic oxidation. In: Spivey JJ, Roberts GW, editors. Catalysis: Volume 17. 17: The Royal Society of Chemistry; 2004. p. 105-51.

[31] Bastos SST, Órfão JJM, Freitas MMA, Pereira MFR, Figueiredo JL. Manganese oxide catalysts synthesized by exotemplating for the total oxidation of ethanol. *Appl Catal B: Environ*. 2009;93(1-2):30-7.

[32] Seiyama T. Total Oxidation of Hydrocarbons on Perovskite Oxides. *Catalysis Reviews*. 1992;34(4):281-300.

[33] Kieûlinga D, Schneidera R, Kraakb P, Haftendornb M, Wendta G. Perovskite-type oxides - catalysts for the total oxidation of chlorinated hydrocarbons. *Appl Catal B: Environ*. 1998;19:143-51.

[34] Merino NA, Barbero BP, Ruiz P, Cadús LE. Synthesis, characterisation, catalytic activity and structural stability of $LaCo1-yFeyO3\pm\lambda$ perovskite catalysts for combustion of ethanol and propane. *J Catal*. 2006;240(2):245-57.

[35] Delimaris D, Ioannides T. VOC oxidation over MnOx–CeO2 catalysts prepared by a combustion method. *Appl Catal B: Environ*. 2008;84(1-2):303-12.

[36] Delimaris D, Ioannides T. VOC oxidation over CuO–CeO2 catalysts prepared by a combustion method. *Appl Catal B: Environ*. 2009;89(1-2):295-302.

[37] Zhang Z, Wan H, Guan G. Catalytic Combustion of Methyl Acetate over Cu-Mn Mixed Oxide Catalyst. *IEEE Xplore*. 2011:1994-7.

[38] Chen X, Carabineiro SAC, Tavares PB, Órfão JJM, Pereira MFR, Figueiredo JL. Catalytic oxidation of ethyl acetate over La-Co and La-Cu oxides. *Journal of Environmental Chemical Engineering*. 2014;2(1):344-55.

[39] Chen X, Carabineiro SAC, Bastos SST, Tavares PB, Órfão JJM, Pereira MFR, et al. Catalytic oxidation of ethyl acetate on cerium-containing mixed oxides. *Appl Catal A: Gen*. 2014;472:101-12.

[40] Suib SL. Microporous manganese oxides. *Current Opinion in Solid State & Materials Science*. 1998;3:63-70.

[41] Brock SL, Duan NG, Tian ZR, Giraldo O, Zhou H, Suib SL. A review of porous manganese oxide materials. *Chemistry of Materials*. 1998;10:2619-28.

[42] Feng Q, Kanoh H, Ooi K. Manganese oxide porous crystals. *Journal of Materials Chemistry*. 1999;9:319-33.

[43] Suib SL. Structure, porosity, and redox in porous manganese oxide octahedral layer and molecular sieve materials. *Journal of Materials Chemistry*. 2008;18:1623-31.

[44] Pasero M. A short outline of the tunnel oxides. Micro and Mesoporous *Mineral Phases*. 2005;57:291-305.

[45] Shen YF, Suib SL, Oyoung CL. Effects of inorganic cation templates on octahedral molecular-sieves of manganese oxide. *Journal of the American Chemical Society* 1994;116:11020-9.

[46] Feng Q, Kanoh H, Miyai Y, Ooi K. Alkali-metal ions insertion/extraction reactions with hollandite-type manganese oxide in the aqueous-phase. *Chemistry of Materials* 1995;7:148-53.

[47] Cai J, Liu J, Willis WS, Suib SL. Framework doping of iron in tunnel structure cryptomelane. *Chemistry of Materials* 2001;13:2413-22.

[48] Chen X, Shen YF, Suib SL, O'Young CL. Catalytic decomposition of 2-propanol over different metal-cation-doped OMS-2 materials. *J Catal.* 2001;197:292-302.

[49] Liu J, Son YC, Cai J, Shen XF, Suib SL, Aindow M. Size control, metal substitution, and catalytic application of cryptomelane nanomaterials prepared using cross-linking reagents. *Chemistry of Materials*. 2004;16(2):276-85.

[50] Liu J, Makwana V, Cai J, Suib SL, Aindow M. Effects of alkali metal and ammonium cation templates on nanofibrous cryptomelane-type manganese oxide octahedral molecular sieves (OMS-2). *The Journal of Physical Chemistry B*. 2003;107:9185-94.

[51] Nicolas-Tolentino E, Tian ZR, Zhou H, Xia GG, Suib SL. Effects of Cu2+ ions on the structure and reactivity of todorokite- and cryptomelane-type manganese oxide octahedral molecular sieves. *Chemistry of Materials*. 1999;11:1733-41.

[52] Liu X-S, Jin Z-N, Lu J-Q, Wang X-X, Luo M-F. Highly active CuO/OMS-2 catalysts for low-temperature CO oxidation. *Chemical Engineering Journal*. 2010;162(1):151-7.

[53] Hernández WY, Centeno MA, Ivanova S, Eloy P, Gaigneaux EM, Odriozola JA. Cu-modified cryptomelane oxide as active catalyst for CO oxidation reactions. *Appl Catal B: Environ*. 2012;123–124(0):27-35.

[54] Ivanova AS, Slavinskaya EM, Stonkus OA, Zaikovskii VI, Danilova IG, Gulyaev RV, et al. Low-temperature oxidation of carbon monoxide over

(Mn1-x M (x))O-2 (M = Co, Pd) catalysts. *Kinetics and Catalysis.* 2013;54(1):81-94.

[55] Gac W. The influence of silver on the structural, redox and catalytic properties of the cryptomelane-type manganese oxides in the low-temperature CO oxidation reaction. *Appl Catal B: Environ.* 2007;75(1-2):107-17.

[56] Hu R, Cheng Y, Xie L, Wang D. Effect of doped Ag on performance of manganese oxide octahedral molecular sieve for CO oxidation. *Chin J Catal.* 2007;28(5):463-8.

[57] Chen J, Li J, Li H, Huang X, Shen W. Facile synthesis of Ag-OMS-2 nanorods and their catalytic applications in CO oxidation. *Microporous and Mesoporous Materials.* 2008;116(1-3):586-92.

[58] Ozacar M, Poyraz AS, Genuino HC, Kuo C-H, Meng Y, Suib SL. Influence of silver on the catalytic properties of the cryptomelane and Ag-hollandite types manganese oxides OMS-2 in the low-temperature CO oxidation. *Appl Catal A: Gen.* 2013;462:64-74.

[59] Liu X, Lu J, Wang X, Luo M. Preparation of Manganese Oxide Octahedral Molecular Sieve and Catalytic Activity of Its Supported PdO for CO Oxidation. *Chin J Catal.* 2010;31(2):181-5.

[60] Martínez T LM, Romero-Sarria F, Hernández WY, Centeno MA, Odriozola JA. Gold supported cryptomelane-type manganese dioxide OMS-2 nanomaterials deposited on AISI 304 stainless steels monoliths for CO oxidation. *Appl Catal A: Gen.* 2012;423–424(0):137-45.

[61] Luo J, Zhang Q, Huang A, Suib SL. Total oxidation of volatile organic compounds with hydrophobic cryptomelane-type octahedral molecular sieves. *Microporours and Mesoporous Materials.* 2000;35-36:209-17.

[62] Tang XF, Huang XM, Shao JJ, Liu JL, Li YG, Xu YD, et al. Synthesis and catalytic performance of manganese oxide octahedral molecular sieve nanorods for formaldehyde oxidation at low temperature. *Chin J Catal.* 2006;27(2):97-9.

[63] Gandhe AR, Rebello JS, Figueiredo JL, Fernandes JB. Manganese oxide OMS-2 as an effective catalyst for total oxidation of ethyl acetate. *Appl Catal B: Environ.* 2007;72(1-2):129-35.

[64] Yu L, Sun M, Yu J, Yu Q, Hao Z, Li C. Synthesis and Characterization of Manganese Oxide Octahedral Molecular Sieve and Its Catalytic Performance for DME Combustion. *Chin J Catal.* 2008;29(11):1127-32.

[65] Chen T, Dou H, Li X, Tang X, Li J, Hao J. Tunnel structure effect of manganese oxides in complete oxidation of formaldehyde. *Microporous and Mesoporous Materials.* 2009;122(1–3):270-4.

[66] Dominguez MI, Navarro P, Romero-Sarria F, Frias D, Cruz SA, Delgado JJ, et al. Fibrous MnO(2) Nanoparticles with (2 x 2) Tunnel Structures. Catalytic Activity in the Total Oxidation of Volatile Organic Compounds. *Journal of Nanoscience and Nanotechnology.* 2009;9(6):3837-42.

[67] Wang R, Li J. Effects of Precursor and Sulfation on OMS-2 Catalyst for Oxidation of Ethanol and Acetaldehyde at Low Temperatures. *Environmental Science & Technology*. 2010;44(11):4282-7.

[68] Yu L, Diao G, Ye F, Sun M, Zhou J, Li Y, et al. Promoting Effect of Ce in Ce/OMS-2 Catalyst for Catalytic Combustion of Dimethyl Ether. *Catal Lett.* 2011;141(1):111-9.

[69] Sun H, Chen S, Wang P, Quan X. Catalytic oxidation of toluene over manganese oxide octahedral molecular sieves (OMS-2) synthesized by different methods. *Chemical Engineering Journal*. 2011;178:191-6.

[70] Sun M, Yu L, Ye F, Diao G, Yu Q, Hao Z, et al. Transition metal doped cryptomelane-type manganese oxide for low-temperature catalytic combustion of dimethyl ether. *Chemical Engineering Journal.* 2013;220:320-7.

[71] Tian H, He J, Zhang X, Zhou L, Wang D. Facile synthesis of porous manganese oxide K-OMS-2 materials and their catalytic activity for formaldehyde oxidation. *Microporous and Mesoporous Materials.* 2011;138(1-3):118-22.

[72] Tian H, He J, Liu L, Wang D. Effects of textural parameters and noble metal loading on the catalytic activity of cryptomelane-type manganese oxides for formaldehyde oxidation. *Ceramics International.* 2013;39(1):315-21.

[73] Hou J, Liu L, Li Y, Mao M, Lv H, Zhao X. Tuning the K+ Concentration in the Tunnel of OMS-2 Nanorods Leads to a Significant Enhancement of the Catalytic Activity for Benzene Oxidation. *Environmental Science & Technology*. 2013;47(23):13730-6.

[74] Wu Y, Yin X, Xing S, Ma Z, Gao Y, Feng L. Synthesis of P-doped mesoporous manganese oxide materials with three-dimensional structures for catalytic oxidation of VOCs. *Materials Letters.* 2013;110:16-9.

[75] Carabineiro SAC, Bastos SST, Órfão JJM, Pereira MFR, Delgado JJ, Figueiredo JL. Carbon Monoxide Oxidation Catalysed by Exotemplated Manganese Oxides. *Catal Lett*. 2010;134(3-4):217-27.

[76] Bowker M, Nuhu A, Soares J. High activity supported gold catalysts by incipient wetness impregnation. *Catal Today*. 2007;122(3-4):245-7.

[77] Carabineiro SAC, Bastos SST, Órfão JJM, Pereira MFR, Delgado JJ, Figueiredo JL. Exotemplated ceria catalysts with gold for CO oxidation. *Appl Catal A: Gen.* 2010;381(1-2):150-60.

[78] Carabineiro SAC, Thompson DT. Catalytic Applications for Gold Nanotechnology. In: Heiz EU, Landman U, editors. Nanocatalysis. Berlin, Heidelberg, New York: Springer-Verlag; 2007. p. 377-489.

[79] Carabineiro SAC, Silva AMT, Dražić G, Tavares PB, Figueiredo JL. Effect of chloride on the sinterization of Au/CeO_2 catalysts. *Catal Today.* 2010;154(3-4):293-302.

[80] Carabineiro SAC, Thompson DT. Gold Catalysis. In: Corti C, Holliday R, editors. Gold: Science and Applications. Boca Raton, London, New York: CRC Press, Taylor and Francis Group; 2010. p. 89-122.

[81] Santos VP, Carabineiro SAC, Tavares PB, Pereira MFR, Órfão JJM, Figueiredo JL. Oxidation of CO, ethanol and toluene over TiO2 supported noble metal catalysts. *Appl Catal B: Environ.* 2010;99(1-2):198-205.

[82] Carabineiro SAC, Bogdanchikova N, Avalos-Borja M, Pestryakov A, Tavares P, Figueiredo JL. Gold supported on metal oxides for carbon monoxide oxidation. *Nano Res.* 2011;4(2):180–93.

[83] Carabineiro SAC, Machado BF, Bacsa RR, Serp P, Dražić G, Faria JL, et al. Catalytic performance of Au/ZnO nanocatalysts for CO oxidation. *J Catal.* 2010;273(2):191–8.

[84] Carabineiro SAC, Silva AMT, Dražić G, Tavares PB, Figueiredo JL. Gold nanoparticles on ceria supports for the oxidation of carbon monoxide. *Catal Today.* 2010;154:21–30.

[85] Cullity BD. Elements of X-Ray Diffraction. Reading, Massachussets: Addison-Wesley Publishing Company, Inc.; 1956.

[86] Christel L, Pierre A, Abel DA-MR. Temperature programmed reduction studies of nickel manganite spinels. *Thermochim Acta.* 1997;306(1–2):51-9.

[87] Wang R, Li J. OMS-2 Catalysts for Formaldehyde Oxidation: Effects of Ce and Pt on Structure and Performance of the Catalysts. *Catal Lett.* 2009;131(3-4):500-5.

[88] Tang X, Li J, Sun L, Hao J. Origination of N2O from NO reduction by NH3 over β-MnO2 and α-Mn2O3. *Appl Catal B: Environ.* 2010;99(1–2):156-62.

[89] Kapteijn F, Vanlangeveld AD, Moulijn JA, Andreini A, Vuurman MA, Turek AM, et al. Alumina-Supported Manganese Oxide Catalysts:

I. Characterization: Effect of Precursor and Loading. *J Catal.* 1994;150(1):94-104.

[90] Nelson AJ, Reynolds JG, Roos JW. Core-level satellites and outer core-level multiplet splitting in Mn model compounds. *Journal of Vacuum Science & Technology a-Vacuum Surfaces and Films*. 2000;18(4):1072-6.

[91] de Vries AH, Hozoi L, Broer R, Bagus PS. Importance of interatomic hole screening in core-level spectroscopy of transition metal oxides: Mn 3s hole states in MnO. *Physical Review B*. 2002;66(3):035108.

[92] Galakhov VR, Demeter M, Bartkowski S, Neumann M, Ovechkina NA, Kurmaev EZ, et al. Mn 3s exchange splitting in mixed-valence manganites. *Physical Review B*. 2002;65(11):113102.

[93] Shen X, Morey AM, Liu J, Ding Y, Cai J, Durand J, et al. Characterization of the Fe-Doped Mixed-Valent Tunnel Structure Manganese Oxide KOMS-2. *The Journal of Physical Chemistry C*. 2011;115(44):21610-9.

[94] Schneider W-D, Laubschat C, Nowik I, Kaindl G. Shake-up excitations and core-hole screening in Eu systems. *Physical Review B*. 1981;24(9):5422-5.

[95] Villegas JC, Garces LJ, Gomez S, Durand JP, Suib SL. Particle Size Control of Cryptomelane Nanomaterials by Use of H2O2 in Acidic Conditions. *Chemistry of Materials*. 2005;17(7):1910-8.

[96] Marschall R, Mukherji A, Tanksale A, Sun C, Smith SC, Wang L, et al. Preparation of new sulfur-doped and sulfur/nitrogen co-doped CsTaWO6 photocatalysts for hydrogen production from water under visible light. *Journal of Materials Chemistry*. 2011;21(24):8871-9.

[97] Qin Y, Lu J, Du P, Chen Z, Ren Y, Wu T, et al. In situ fabrication of porous-carbon-supported [small alpha]-MnO2 nanorods at room temperature: application for rechargeable Li-O2 batteries. *Energy & Environmental Science*. 2013;6(2):519-31.

[98] Shyu JZ, Otto K, Watkins WLH, Graham GW, Belitz RK, Gandhi HS. Characterization of Pd/γ-alumina catalysts containing ceria. *J Catal.* 1988;114(1):23-33.

[99] Scirè S, Minicò S, Crisafulli C, Satriano C, Pistone A. Catalytic combustion of volatile organic compounds on gold/cerium oxide catalysts. *Appl Catal B: Environ*. 2003;40(1):43-9.

In: Volatile Organic Compounds
Editor: Julian Patrick Moore
ISBN: 978-1-63485-370-5

Chapter 6

VOLATILE ORGANIC COMPOUNDS FROM TRUFFLES AND FALSE TRUFFLES FROM BASILICATA (SOUTHERN ITALY)

Maurizio D'Auria, Gian Luigi Rana and Rocco Racioppi
Dipartimento di Scienze, Università della Basilicata,
Potenza, Italy

ABSTRACT

Volatile organic compounds (VOCs) of several Tuber species are identified via solid-phase microextraction-gas-chromatography-mass spectrometry analysis. The VOCs of *T. mesentericum*, *T. exacavatum*, *T. borchii*, *T. magnatum*, *T. aestivm*, *T. uncinatum*, *T. brumale*, *T. melanosporum*, *T. oligospermum*, *T. panniferum* have been determined.

Ascomata of two truffle species, *Tuber borchii* and *T. asa-foetida* were identified on the base of ascospore morphology and compared under volatile organic compound profile to determine the particular VOCs which characterize each taxon. SPME-GC-MS analysis of the samples showed the presence of 1-methyl-1,3-butadiene as a main component in both the truffles. *T. asa-foetida* showed a compound, toluene, not present in *T. borchii*, in agreement with a penetrant "solvent" smell of the truffle.

Volatile organic compounds (VOCs) of *Schenella pityophilus* have been identified via solid-phase microextraction-gas chromatography-mass spectrometry analysis. Ten compounds have been identified. 3-Methylthio-1-propene was the most significant component. Some other components were identified previously in *Tuber aestivum* and *Tuber melanosporum*.

Results of SPME-CG-MS analyses, accomplished on sporophores of eleven species of truffles and false truffles, are reported. VOCs found in *Gautieria morchelliformis* were dimethyl sulphide, 1,3-octadiene, 3,7-dimethyl-1,6-octadien-3-ol, γ-muurolene, amorphadiene, isoledene, and *cis*-muurola-3,5-diene. In *Hymenogaster luteus* var. *luteus,* presence of 1,3-octadiene, 1-octen-3-ol, 3-octanone, 3-octanol, and 4-acetylanisole was revealed. Two VOCs, 4-acetylanisole and β-farnesene, constituted aroma of *Hymenogaster olivaceus. Melanogaster broomeanus* exhibited as components of its aroma 2-methyl-1,3-butadiene, 2-methylpropanal, 2-methylpropanol, isobutyl acetate, 3,7-dimethyl-1,6-octadien-3-ol, 3-octanone, and β-curcumene. *Melanogaster variegatus* showed the presence of 2-methylpropanol, ethyl 2-methylpropanoate, isobutyl acetale, 2-methypropyl 2-methyl-2-butenoate, 3-phenylpropyl acetate, 2-methylpropyl propanoate. VOC profile of *Octavianina asterosperma* was characterized by the presence of dimethyl sulphide, ethyl 2-methylpropanoate, methyl 2-methylbutanoate, and 3-octanone. *Choiromyces meandriformis*, *Tuber rufum* var. *rufum*, *T. rufum* var. *lucidum* and *Pachyphloeus conglomeratus* showed the presence of dimethyl sulphide only. Some methoxy substitued ethers were, finally, present in *Tuber dryophilum*.

INTRODUCTION

The odour is one of the main general characteristics of truffles (*Ascomycetes*) and false truffles (*Basidiomycetes*) that not rarely make easy to recognize their different species. On the other hand, the description of the above scent is sometimes rather vague or can be source of uncertainty (Montecchi and Sarasini, 2000). Examples of the above incongruity, regarding sporophores of four false truffles are hereafter reported:

a. *Hymenogaster luteus* Vittad. var *luteus*. They would have, according to Vittadini (1831), a strawberry smell which *vice versa* is described by other authors (Montecchi and Sarasini, 2000) as very like fig leaves or, even, lavender;
b. *H. muticus* Berkeley and Boome (1848). These would have a weak (or slender), not bad scent (which however remains indefinite) (Montecchi and Sarasini, 2000);

c. *Gautieria morchelliformis* Vittad. Kuntze (Vittadini 1831) would be initially characterized by a pleasant smell of exotic fruits (without any specific detail). The same odour would then become nauseous.
d. *Octavianina asterosperma* (Vittad.) Kuntze would have a fruit or sweetmeat (i.e., candy like) smell (Vittadini, 1831).

Although studies on biodiversity of hypogeous fungi naturally growing in Basilicata (southern Italy) begun only 15 years ago, the number of truffles and false truffles so far discovered in the region nowadays comprehends about 65 taxa and puts Basilicata at the 4th position in the list of Italian regions which carry out researches in this interesting field of mycology (Rana et al., 2008; Rana et al., 2010; Cerone et al., 2004).

Among truffle species which grow and are collected in the region, *T. borchii* Vittad. follows only *T. aestivum* Vittad. under quantitative profile and *T. magnatum* Pico for qualitative features. In fact, its natural beds range from sea level to mountain (Cerone et al., 2004)) and its aroma is sweetish, of truffle, with a weak alliaceous component.

Its habitat is constituted by mixed forests or conifers and it is usually collected from October to April. Actually, *T. borchii* is considered as a species complex comprehending at least 5 other truffles, i.e., *T. gibbosum* Harkness, *T. puberulum* Berk & Br., *T. maculatum* Vittad., *T. foetidum* Vittad., and *T. dryophilum* Tul. & C. Tul., which have maturation periods and habitats or symbiotic plants (except for *T. gibbosum*) almost overlapping that of *T. borchii* (Montecchi and Sarasini, 2000).

During the last thirty years the volatile organic compounds (VOCs) in truffle aroma have been analysed by using several methods (Fiecchi et al., 1967; Ney and Frietag, 1980; Claus et al., 1981; Balestreri et al., 1986; Talou et al., 1987; Bellina Agostinone et al., 1987; Angeletti et al., 1988; Fiecchi, 1988; Balestreri et al., 1988; Hanssen and Kühne, 1988; Flament et al., 1990; Pacioni et al., 1990; Talou et al., 1990 ; Bellesia et al., 1996a ; Bellesia et al., 1996b ; Bellesia et al., 1998a; Bellesia et al., 1998b). Black truffle (*Tuber melanosporum* Vitt.) has been extensively studied and several compounds have been identified in its flavour. On the contrary, bis(methylthio)methane was identified as the main component of white truffle (*Tuber magnatum* Pico) aroma (Fiecchi et al., 1967). VOCs have been determined also in the mycelium of *Tuber borchii* Vitt. few years ago (Tirillini et al., 2000; Bellesia et al., 2001).

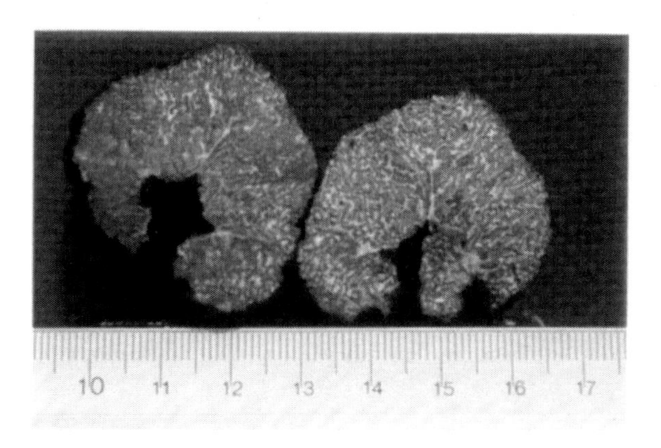

Figure 1. *Tuber mesentericum.*

SPME was applied to the analysis of flavours (Boyd-Boland et al., 1994; Yang and Peppard, 1994; Chin et al., 1996; Matich et al., 1996; Clark and Bunch, 1997; Elmore et al., 1997; Song et al., 1997; Steffen and Pawliszyn, 1997). The use of SPME in the determination of sulphur components of black and white truffle flavour has been also reported (Pelusio et al., 1995).

Results

Tuber Mesentericum

Twenty-seven VOCs were identified in samples of *T. mesentericum* (Figure 1, Table 1). The same table contains the per cent area for each identified component.

This is the first report dealing with the aroma of this truffle species. The compounds found with high frequency were dimethylsulphide (found on 100% of the samples), 3-methylanisole (79%), 2,5-dimethoxytoluene (57%), 3-methylbutanal (43%), and butanone (36%).

Tuber Excavatum

VOCs found in *T. excavatum* (Figure 2) are listed in Table 2. High result variability was observed in the flavour of analysed samples. Dimethylsulphide and 3-octanone were present in three of the four ascocarps subjected to analysis. Other VOCs were found only in a couple of samples, i.e., 3-octanol, ethanol, acetic acid, and decane, whereas the remaining ones were encountered in single carpophores.

Tuber Borchii

Samples of *T. borchii* (Figure 3) also showed a high variability in the aroma composition (Table 3). It is noteworthy that all of them were lacking dimethylsulphide. Two VOCs (2-methyl-1,3-butadiene and 1,2-pentadiene) were found with high per cent area in two ascocarps which also contained 5-11 other volatile substances. VOC number per sample varied from 6 to 16. Sometime 1,2-pentadiene was present in lower percentage but was accompanied by discrete amounts of 1-methylpropyl formate, tetradecanal, and tetradecane, or by 3-octanone and very little amounts of other 15 VOCs.

Tirillini and coworkers analysed VOCs in mycelium of *T. borchii* (Tirillini et al., 2000). We have to note that there is not superposition between our results and those reported in that work. In fact, only 3-octanone, decane, and ethynylbenzene were found in both the studies.

Figure 2. *Tuber excavatum.*

Table 1. VOCs identified in *T. mesentericum* and corresponding per cent area

Compound	Sample													
	1	**2**	**3**	**4**	**5**	**6**	**7**	**8**	**9**	**10**	**11**	**12**	**13**	**14**
1,2-pentadiene		1.1	0.4											1.3
2-methyl-1,3-butadiene	0.8			0.2		2.6		2.4						
thiourea					0.2									
dimethylsulphide	3.4	0.1	1.9	4.5	1.8	5.4	20.0	0.3	3.3	4.1	1.3	1.3	2.9	10.2
1-propanol					1.8									
2-butanol					2.5									
butanone	3.5	2.2								8.0			3.0	13.4
2-methylbutanal									6.4			3.4	0.4	1.1
3-methylbutanal			2.8						29.6		1.5	22.7	8.2	4.8
hexanal													0.3	0.1
1,2-propandiol					0.8									
2,3-butandiol					2.5									
heptanal													0.1	
3-octanone											6.7		0.8	
benzaldehyde													0.3	0.6
1-methoxy-3-methylbenzene	1.4	80.1	7.7	72.8	29.3	65.0	21.6	83.0			6.4		0.9	0.7
1-undecene					0.7									
2-ethylanisole				0.7				0.2						
2-phenylethanol					0.3		0.8							
2,3-dimethoxytoluene		0.3		0.1	0.2			0.3						
nonanol					0.2									
2,5-dimethoxytoluene		6.1		2.4	1.5		0.5	8.1			12.4		0.3	1.0
3,4-dimethoxytoluene						0.3							0.3	
methyldiethyldithiocarbamide				0.1										
2,3,5-trimethoxytoluene				0.1										1.0
hexacosane		0.8												
heptacosane		0.9												

Table 2. VOCs identified in *T. excavatum* and corresponding per cent area

Compound	Sample			
	1	2	3	4
ethanol		38.9	2.3	
dimethylsulphide		2.4	3.4	2.2
2-methyl-1-propanol			4.4	
acetic acid		4.5	0.4	
3-methylbutanal	4.8			
cis-methylpropenylsulphide				1.5
3-methyl-1-butanol			5.1	
2-methyl-1-butanol			2.8	
toluene			0.3	
1,3-dimethylbenzene			0.1	
styrene			0.3	
3-octanone	6.4	5.2	17.4	
benzaldehyde	7.2			
1-octen-3-ol	48.5			
2-pentylfuran	0.7			
3-octanol	6.2		31.3	
decane		1.1	1.3	
benzo[c]thiophene			0.2	

Figure 3. *Tuber borchii.*

Table 3. VOCs identified in *T. borchii* and corresponding per cent area

Compound	Sample				
	1	2	3	4	5
ethanol	0.7		0.8		
1,2-pentadiene	63.1	5.7		6.4	
2-methyl-1,3-butadiene			42.7		2.3
2-butanol				0.1	
2-methylfuran	0.2		2.5	0.2	
2-methyl-1-propanol			0.2		
2-methylbutanal		0.7	0.6	0.3	
3-methylbutanal	0.1		2.8		
pentanal				1.0	
1-methylpropyl formate		8.6			
3-methyl-1-butanol			2.3		
2-methyl-1-butanol		1.8	0.6	0.5	
toluene		0.3		0.1	
3-methylthiophene	0.3		0.3	0.3	
1,3-dimethylbenzene		0.2		0.1	
styrene		0.2		0.1	
3-octanone			0.6	6.6	
β-ocimene	0.7		0.3	1.2	
decane		1.7	0.5	0.9	
benzo[c]thiophene		0.2		0.4	
tetradecane		23.2		0.1	
tetradecanal		17.2		1.2	

Tuber Asa-Foetida

Another species of *Tuber* which naturally grows in Basilicata and can be macroscopically confused with *T. borchii* is Tul. & C. Tul., although it namely grows in symbiosis with various species of *Cistus*, *Helianthemum*, *Ephedra* and other *Cistaceae* in sandy soils close to sea, has a strong, mouldy and nauseating odour and generally matures after *T. borchii*.

Ascospores of *T. borchii* and *T. asa-foetida* were easily distinguished each from the other. In fact, in the first case, they appeared in great majority ellipsoid, reticulate-alveolate with more than 7-8 poligonal meshes along the longer axis, but also subglobose, and variable in dimension (25-40 μm on average) but not rarely reaching 25,6 x 48μm (Figure 4a, b).

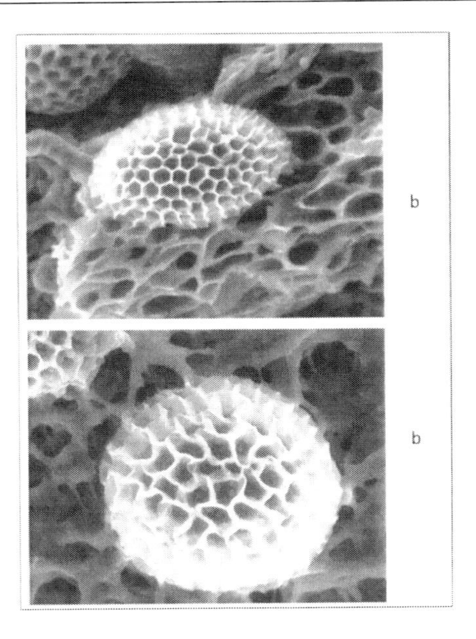

Figure 4. ESEM of *Tuber borchii*.

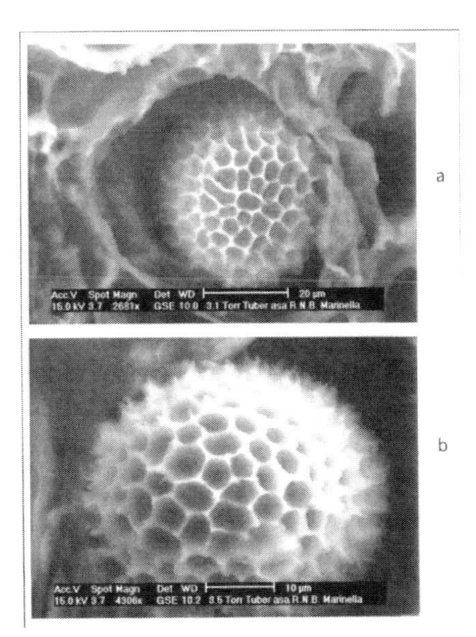

Figure 5. ESEM of *Tuber asa-foetida*.

Ascospores of the second fungus species were, vice versa, ellipsoid-subglobose, but also spherical, reticulate-alveolate with 6-8 meshes along their diameter and had dimensions averaging from 25-30 μm up to 35-45μm (Figure 5a, b).

We analyzed three samples of *T. Borchii* and the results are reported in Table 4. In a previous work in this field, we found that all samples lacked dimethylsulfide and that 2-methyl-1,3-butadiene and 1,2-pentadiene were the main components of the volatile fractions (see above). Furthermore, lower percentage of 1-methylpropyl formate, tetradecanal, tetradecane, and 3-octanone were found. In other work on this tuber both 2-methyl-1,3-butadiene and 1,2-pentadiene have not been found, while 1-octen-3-ol was found to be the main component of the VOCs. Furthermore, some sulfur compounds was found (Bellesia et al., 2001; Zeppa et al., 2004; Menotta et al., 2004). In this study, the main component of the VOCs in *T. borchii* is 2-methyl-1,3-butadiene, while the other components are present in very low amount. Minor components were 3-methylbutanal, 3-methyl-1-butanol, and tetradecane.

We analyzed four samples of *T. asa-foetida* at different maturation level (from immature, sample 1, until very mature, sample 4) and the results are reported in Table 4. In this type of tuber, the amount of 2-methyl-1,3-butadiene is lower than in *T. borchii*, while we observe the presence of other volatile compounds in relevant amounts such as 2-butanone, 2-methyl-1-propanol, 2-methyl-1-butanol, and toluene. Probably, the latter compound is responsible of the "solvent" smell of this tuber.

In the case of 2-methyl-1-propanol, 3- and 2-methyl-1-butanol we were able to give a quantitative evaluation of the concentration of these compounds in the tubers. The results are reported in Table 5.

Tube Aestivum

Analyses of carpophores of *Tuber aestivum* (Figure 6) and *T. aestivum* f. *uncinatum* gave the results showed in Table 6. Dimethysulphide was the compound found with the highest frequency (85% of the samples). Butanone and 3-methylbutanal were also identified in 60% and 55% of specimens, respectively. Other VOCs found in discrete amounts in some samples were dimethylsulphide, 1-methoxy-3-methylbenzene, butanone, ethanol, and ethyl acetate. It is noteworthy that in *T. aestivum* f. *uncinatum* the amounts of dimethylsuphide were always lower than in *T. aestivum*.

Table 4. VOCs identified in *T. borchii* and *T. asa-foetida*

Compound	r.t. [min.]	Area %						
		T. borchii			*T. asa-foetida*			
		1	2	3	1	2	3	4
2-methyl-1,3-butadiene	1.63	18.1	50.3	59.9	17.3	17.2	23.8	44.9
2-butanone	2.07							3.9
2-methylfuran	2.09		0.3	0.3		0.6		0.8
2-methyl-1-propanol	2.16		0.3			8.1		5.2
tetrahydrofuran	2.26				3.5		1.2	
3-methylbutanal	2.30	0.4	1.5			0.3		
benzene	2.49				0.6		0.8	
1-methylpropyl formate	2.64							0.2
3-methyl-1-butanol	3.30	0.9	1.8			16.9		0.4
2-methyl-1-butanol	3.32							6.1
toluene	3.81				2.1	1.1	1.2	0.2
3-methylthiophene	4.06	0.9	0.8	0.5				
xylene	5.80		0.1	0.3				
α-pinene	7.21		0.1					
3,7-dimethyl-1,3,6-octatriene	9.78		0.2					
tetradecane	16.27	0.4	0.2	0.2		1.5		
3-acetyl-1-propyl-5,6-dihydro-2-naphthol	17.39		0.1					
9-(diphenylmethylene)-9H-fluorene	29.88	1.9						

Table 5. Concentration of selected VOCs identified in *T. borchii* and *T. asa-foetida*

Compound	Concentration [mg g^{-1}]						
	Tuber borchii			*Tuber asa-foetida*			
	1	2	3	1	2	3	4
2-methyl-1-propanol		0.029			0.072		0.263
3-methyl-1-butanol	0.170	0.322			1.139		0.046
2-methyl-1-butanol							0.837

The above results are in substantial agreement with those of Bellina Agostinone (1987) who found same main VOCs in both the truffle entities. On the contrary, other VOCs (i.e., 3- and 2-methylbutanal, 2-phenylethanol, 3- and 2-methyl-1-butanol) found by Bellesia et al., (1998a) in *T. uncinatum* were not detected here.

Figure 6. *Tuber aestivum*.

Tuber Magnatum

SPME-GC-MS analysis of white truffle samples (*T. magnatum*) showed that only seven VOCs were present (Figure 7, Table 7). Dimethylsulphide was present in all the examined samples in percentage varying from 0.4 to 16.6%. With the exception of 2- and 3-methylbutanal (found in two samples), only sulphur compounds were present. The most abundant VOC, in agreement with the results reported by Hansen and Kühne (1987) and Bellesia et al., (1996b), was 2,4-dithiopentane.

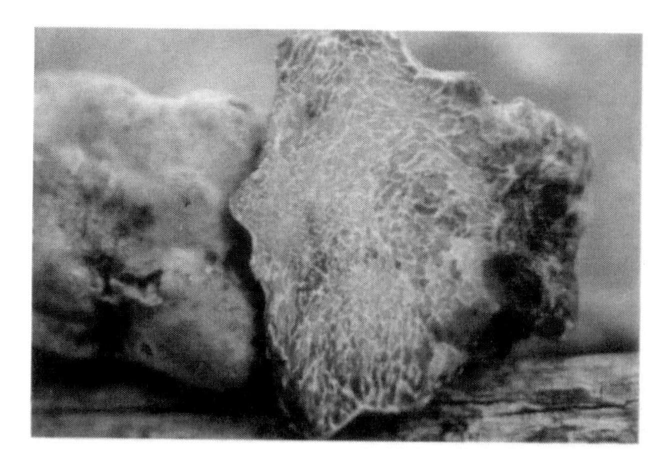

Figure 7. *Tuber magnatum*.

Table 6. VOCs identified in samples of *Tuber aestivum* and its form *uncinatum* (*) and corresponding per cent area

Compound	Sample																			
	1	2 (*)	3 (*)	4	5	6	7	8	9	10	11	12	13	14	15	16	17	18	19	20
ethanol	0.4			9.3	22.6	22.9				1.9										0.1
propanone	0.6							0.8												
dimethylsulphide	7.9	2.2	2.9	24.8		5.2	4.0	5.5		8.0	6.9		11.3	5.5	3.7	24.1	16.1	16.2	10.4	6.3
1-propanol				2.1																
2-methylpropanal							1.0	1.5												
2-butanol				2.6																
butanone	54.5	2.3	18.9					3.3		4.0	2.2		37.7			6.9	2.5	10.7	4.8	6.2
ethyl acetate	0.1				2.6													47.4	15.9	
2-methyl-1-propanol	0.9			2.4	3.4															
2-methylbutanal							0.9	4.3	1.2			1.1						0.5	0.9	0.1
3-methylbutanal	0.3						1.9	9.4	3.1	0.9	1.7	3.8		1.1				0.4	1.5	0.7
1-methylpropyl formate	0.4																			
3-methyl-1-butanol				2.1	3.8			1.2		1.2										
2-methyl-1-butanol	1.1			7.9	3.9	3.5														
1-methylpropyl acetate	0.5																			
toluene			0.4																	
1,3-butandiol					1.4															
2,3-butandiol					1.4					0.5										
3-octanone					0.8															0.4
benzaldehyde								1.0												
2-pentylfuran								0.7												
1-methoxy-3-methylbenzene		1.4	9.6				28.7	0.5		5.2										8.0
2-phenylethanol					1.3															

Table 7. VOCs identified in *T. magnatum* and corresponding per cent area

Compound	Sample				
	1	2	3	4	5
dimethylsulphide	16.6	0.4	9.4	0.7	3.8
2-methylbutanal		0.6	0.1		
3-methylbutanal		1.7	0.1		
dimethyldisulphide	0.6		0.3		
2,4-dithiopentane			57.0	2.1	6.4
dimethyltrisulphide			0.1		
trimethyltrisulphide			1.2		

Figure 8. *Tuber brumale*.

We did not find 1,2,4-trithiolane, methyl(methylthio)methyldisulphide, and tris(methylthio)methane as reported by Pelusio et al. (1995).

Tuber Brumale

Twenty-eight VOCs were found in *T. brumale* (Figure 8) and *T. brumale* f. *moschatum* (Table 8). The most representative compounds found in *T. brumale* were dimethylsulphide, butanone, 2-methylbutanal, 1-methylpropyl formate, 2-methylpropanal, and 1,4-dimethoxybenzene. In *T. brumale* f. *moschatum,* dimethylsulphide reached lower amounts than in *T. brumale*. The main other VOCs found were 2-methylpropanal, butanone, 2-methylbutanal, and 1,4-dimethoxybenzene. It is interesting to note that our data fit the result reported by Bellesia (1996a). Furthermore, in this study six compounds never

FALSE TRUFFLE

Schenella Pityophilus

Schenella pityophilus Malençon and Riousset (Malençon and Riousset, 1977; Estrada-Torres et al., 2005) is hypogeous basidiomycete, a false truffle, belonging to *Agaricomycetes, Phallomycetidae, Geastrales, Geastraceae* (Hibbet et al., 2007), previously classified as *Pyrenogaster pityophilus* (Malençon and Riousset, 1977) (Figure 11).

Table 12. VOCs identified in some other Ascomycetes and corresponding per cent area

Compound	Retention time [min.]	*Choroimyces meandriformis*	*Tuber rufum* var. *rufum*	*Tuber rufum* var. *lucidum*	*Tuber dryophilum*	*P. conglomeratus*
Dimethyl sulfide	1.70	18.95	56.70	6.53		
2-Methylpropanol	2.25			2.72		
3-Methylbutanal	2.50			1.29		
1-Methoxy-2-methyl-3-butene	2.61				0.90	
Methyl 2-methylpropanoate	2.80					
1-Methoxy-3-methylbutane	2.82			1.88	2.35	
3-Methylbutanol	3.57			1.41		
Dimethyldisulfide	3.70					18.46
Toluene	4.07					1.01
Cis-1-methoxy-3-methyl-1-butene	5.55				1.15	
1-Methoxy-3-methylbenzene	9.25		0.63			
D-limonene	9.39		0.28			
1-Undecene	10.61	0.19	0.25			0.80
2-Nonanone	10.75					1.27
2-Undecanone	14.43					1.25
1,3-diisopropylnaphthalene	20.44			0.06		
1,7-diisopropylnaphthalene	20.52			0.18		
2,6-diisoprpylnaphthalene	21.14			0.07		

Although it is considered in literature as a rare species in Europe (Malençon and Riousset, 1977; Montecchi and Sarasini, 2000; Venturella et al., 2004), some years ago it was found in three pinewoods in Salento (Southern Italy) (Signore et al., 2008) and, more recently, in two different places in Basilicata (Southern Italy). The complete description of the fungus, showing interesting and curious morphological characters, has been performed (Malençon and Riousset, 1977; Poumarat and Neville, 1997; Montecchi and Sarasini, 2000; Venturella et al., 2004; Signore et al., 2008). This false truffle emanates a pleasant smell, similar to that of *Tuber aestivum* Vittad. fo. uncinatum (Chatin) or to that of *Tuber magnatum* Pico (Montecchi and Borelli, 1990).

Figure 11. *Schenella pityophilus*.

Table 13. VOCs from *Schenella pityophilus*

Compound	t_R [min]	Area in young fungus	Area % in young fungus	Area in mature fungus	Area % in mature fungus
Dimethylsulfide	1.81			1241279	1.42
2-Methylpropanal	1.92			2513187	2.87
2-Methylpropanol	2.21	807428	0.47	802568	0.92
3-Methylthio-1-propene	2.79	3431767	2.85	1213737	1.24
Octene	4.19			120441	0.13
1-Octen-3-ol	8.35	351973	0.32		
3-Octanone	8.48	1276221	1.17	526660	0.60
4,6-Dimethyldodecane	13.53			165532	0.19
m-Di-*t*-butylbenzene	13.78			272510	0.32
Squalene	29.55	990266	0.91		

Table 13 collects the area percent found for all the compounds we determined in the fungus. We did not perform a quantitative analysis. We were able to estimate the amount of dimethylsulfide (0.02 μg), and the concentration of 2-methylpropanol (0.14 μg/g). The young fungus showed only five compounds where the main components were 3-methylthio-1-propene and 3-octanone. The mature fungus, on the other side, showed the presence of eight compounds, where the main components were dimethysulfide, 2-methylpropanal, and 3-methylthio-1-propene. Furthermore, 2-methylpropanol and 3-octanone were minor components. Then, this fungus, such as truffles, showed sulfur containing compounds as the main components of the aroma.

Table 14. VOCs identified in *G. morchelliformis* and corresponding per cent area

Compound	Retention time [min.]	Area %
Dimethyl sulfide	1.70	5.56
3-Methylbutanal	2.50	0.22
Dimethyldisulfide	3.70	0.07
3-Methyl-1, 3, 5-hexatriene	4.03	0.06
1-Octene	4.45	0.68
1,3-Octadiene	5.12	6.92
2-Methylnonane	8.01	0.10
3-Octanone	8.51	0.15
4-Methyldecane	9.22	0.19
D-limonene	9.39	0.29
Trans-β-Ocimene	9.58	0.07
Cis-Linaloloxide	10.30	0.02
4-Carene	10.59	0.33
3,7-Dimethyl-1,6-octadien-3-ol	10.82	1.03
1-Octen-3-yl acetate	11.05	0.18
α–Terpineol	12.63	0.15
α-Copaene	15.88	0.24
Epizoranene	16.32	0.12
γ-Muurolene	16.67	1.33
Epi-biscyclosesquiphellandrene	16.77	0.56
Amorphadiene	17.27	40.53
Isoledene	17.85	3.75
β-Bisabolene	17.93	0.81
Cis-muurola-3,5-diene	18.21	5.19
β-Vativerene	18.87	0.12

It is noteworthy that 3-octanone was found in *T. mesentericum, T. excavatum, T. borchii, T. aestivum,* and *T. brumale*, 2-methylpropanal was found in *T. aestivum, T. brumale,* and *T. melanosporum*, while 2-

methylpropanol was found in *T. excavatum, T. borchii, T. aestivum, T. brumale,* and *T. melanosporum* (see above). These minor components (3-octanone, 2-methylpropanol, and 2-methylpropanal) are similar to those found mainly in *T. aestivum* and *T. melanosporum.* One of the main components we found in *Schenella pityophilus* was 3-methylthio-1-propene. In our knowledge this is the first discovery of this compound in a fungus, while it was found in garlic and onion (Yu et al., 1989; Shaath and Flores, 1998; Kim et al., 2004; Calvo-Gomez et al., 2004).

Gautiera Morchelliformis

In *G. morchelliformis*, dimethyl sulphide, 1,3-octadiene, 3,7-dimethyl-1,6-octadien-3-ol, γ-muurolene, amorphadiene, isoledene, and *cis*-muurola-3,5-diene have been found (Table 14). The main component of *G. morchelliformis* aromawas amorphadiene (Komatsu et al., 2010) which is an important precursor of the antimalarial drug artemisinin. The aroma of this species is probably a mixture of dimethyl sulphide and terpenes which confere the enough pleasant camphoric smell which could be like that of exotic fruit (i.e., immature mango fruit) reported in literature.

Hymenogaster Luteus

H. luteus var. *luteus* showed the presence of 1,3-octadiene, 1-octen-3-ol, 3-octanone, 3-octanol, and 4-acetylanisole (Table 15). ESEM is reported in Figure 12.

The main components of the aroma of *H. luteus* resulted 1,3-octadiene (24.78%) and 3-octanone (14.48%) whereas 1-octen-3-ol (4.89%) is a minor constituent. 1,3-Octadiene is usually associated to a fungal aroma while 1-octen-3-ol, and 3-octanone are responsible of a cheesy odour. 4-Acetylanisole, another component of scent of this false truffle, exhibits a butter, caramel, fruity, vanilla aroma. Mixture of the above VOCs in *H. luteus* var. *luteus*, could in part explain why description of its scent in previous literature results quite complex and variable from mushroom-like to those of strawberry, fig leaves and even lavender.

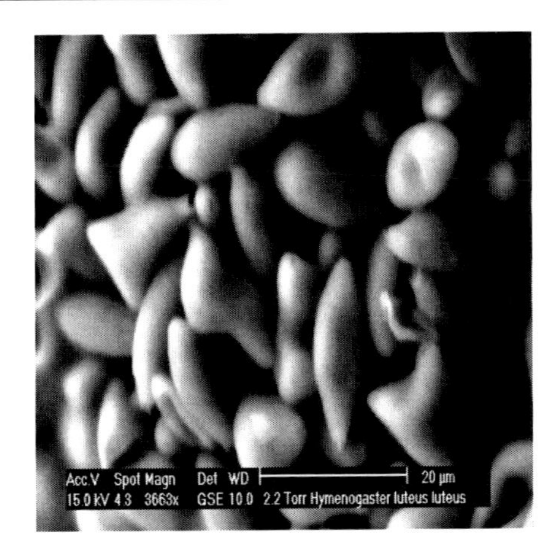

Figure 12. Basidiospore di *Hymenogaster luteus* var. *luteus*.

Table 15. VOCs identified in *Hymenogaster luteus* var. *luteus* and corresponding per cent area

Compound	Retention time [min.]	Area %
1,3-Octadiene	5.12	24.78
Xylene	6.08	0.11
2-*n*-Butylfuran	6.55	0.29
1-Octen-3-ol	8.38	4.89
3-Octanone	8.51	14.48
3-Octanol	8.71	7.20
Eucalyptol	9.46	0.03
(*E*)-1,4-Undecadiene	10.54	0.10
1-Undecene	10.61	1.46
1-Octen-3-yl acetate	11.05	0.20
1,4-Dimethoxybenzene	12.27	0.21
Methyl 3,5-dimethylbenzoate	15.45	0.11
1-Methylethyl dodecanoate	19.62	0.04
1-Butyloctylbenzene	21.15	0.03

Hymemogaster Olivaceus

In *H. olivaceus* Vittad., anisole, 1-methoxyethylbenzene, 1-undecene, 4-acetyl-1-methylcyclohexene, 1-ethenyl-4-methoxybenzene, 4-acetylanisole, 1,2,4-trimethoxybenzene, γ-muurolene, neoisologifolene, β-curcumene,β-farnesene, *cis*-α-bisabolene, and β-bisabolene have been detected (Table 16).

H. olivaceus showed as main aroma components 4-acetylanisole (4.48%) and β-farnesene (11.44%). The aroma of 4-acetylanisole and the woody odour of β-farnesene are, in mixture, perfectly compatible with the "switish, of fruit compote, or rather sour of woody essences" aroma already reported for its ascomata (Montecchi & Sarasini, 2000).

Table 16. VOCs identified in *Hymenogaster olivaceusand* corresponding per cent area

Compound	Retention time [min.]	Area %
1-Octene	4.45	0.10
Anisole	7.09	0.60
3-Octanone	8.51	0.25
1-Mehoxyethylbenzene	8.91	1.56
Acetophenone	10.32	0.15
1-Undecene	10.61	0.84
6-Methyl-3,5-heptadiene-2-one	11.01	0.57
4-Acetyl-1-methylcyclohexene	11.46	0.87
1,2-Dimethoxybenzene	11.76	0.21
1-Ethenyl-4-methoxybenzene	11.90	0.90
1,4-Dimethoxybenzene	12.27	0.48
α–Cubebene	15.40	0.16
4-Acetylanisole	15.52	8.48
1,2,4-Trimethoxybenzene	15.77	1.70
γ-Muurolene	16.67	0.61
Neoisolongifolene	16.87	0.90
Humulene	17.16	0.13
β-Curcumene	17.36	0.41
β-Farnesene	17.61	11.44
Cis-α-Bisabolene	17.83	0.91
β-Bisabolene	17.93	0.52

Melanogaster Broomeanus

M. broomeanus Berk. *apud* Tul. & C. Tul. has as volatile organic compounds 2-methyl-1,3-butadiene, 2-methylpropanal, isobutyl acetate, 2-methylpropanol, 3-octanone, 3,7-dimethyl-1,6-octafien-3-ol, octyl acetate, γ-muurolene, 1-methyl-4-(1-methylethyl)-1,3-cyclohexadiene, and β-curcumene (Table 17). The second main component of *M. broomeanus* aroma (2-

methylpropanal) shows a fresh aldehydic floral odour. Also isobutyl acetate has a fruity and floral smell. Furthermore, 3,7-dimethyl-1,6-octadien-3-ol is known for its floral, sweet, woody odour whereas β-curcumene is not known for its use as a fragrance due to its "peppery-saffron-sulphuric" scent. All the above VOCs could explain why aroma of this *Melanogaster* species is reported in literature as "weak, pleasant, aromatic and fruity" (Montecchi & Sarasini, 2000).

Table 17. VOCs identified in *Melanogaster broomeanus* and corresponding per cent area

Compound	Retention time [min.]	Area %
2-Methyl-1,3-butadiene	1.67	4.09
2-Methylpropanal	1.84	6.15
2-Methylpropanol	2.25	3.49
Ethyl propanoate	3.17	0.80
Isobutyl acetate	4.19	8.35
2-Methylpropyl propanoate	6.99	0.10
3-Octanone	8.51	6.35
Trans-β-Ocimene	9.58	0.08
Cis-Linaloloxide	10.30	0.15
3,7-Dimethyl-1,6-octadien-3-ol	10.82	6.62
Octyl acetate	12.89	1.00
2,6-Dimethyl-2,6-octadiene	15.42	0.05
Aromadendrene	15.92	0.17
Diepi-α-cedrene	16.01	0.11
Longifolene	16.41	0.20
γ-Muurolene	16.67	0.77
Cis-Thujopsene	16.82	0.33
1-Methyl-4-(1-methylethyl)-1,3-cyclohexadiene	16.90	0.80
β-Curcumene	17.36	3.06
Cis-α-Bisabolene	17.83	0.12
Hexadecane	19.19	0.12
Heptadecane	20.61	0.06
2,6,10,14-Tetramethylpentadecane	20.68	0.06

Melanogaster Variegatus

The GC-MS spectrum of *M. variegatus* (Vittad.) Tul. & C. Tul., showed that in this species the volatile organic compounds were 2-methyl-1,3-butadiene, 2-methylpropanal, 2-methylpropanol, ethyl 2-methylpropanoate, isobutyl acetate, 2-methypropyl propanoate, 2-methylpropyl 2-methyl-2-butenoate, 3-phenylpropyl acetate, and 2-methylpropyl propanoate (Table 18).

Table 18. VOCs identified in *Melanogaster variegatus* and corresponding per cent area

Compound	Retention time [min.]	Area %
2-Methyl-1,3-butadiene	1.67	2.11
2-Methylpropanal	1.84	2.33
2-Methylpropanol	2.25	4.32
Ethyl propanoate	3.17	0.69
Ethyl 2-methylpropanoate	3.91	3.28
Isobutyl acetate	4.19	14.11
Ethyl methacrylate	4.39	1.58
Propyl propanoate	4.85	0.34
Ethyl 2-butenoate	5.58	0.10
3-Methylbutyl acetate	6.02	0.82
2-Methylpropyl propanoate	6.99	7.13
2-Methylpropyl 2-methylpropanoate	7.51	1.61
2-Methylpropyl butanoate	7.85	0.16
Isopropyl tiglate	8.27	0.06
3-Octanone	8.51	1.11
Propyl 2-methyl-2-butenoate	9.54	0.15
2-Methylpropyl 2-methyl2-butenoate	10.66	3.02
Benzenepropanal	12.20	0.15
Octyl acetate	12.89	0.92
3-Phenyl-2-propenal	14.23	0.14
1,2-Dihydro-3-methylnaphthalene	14.53	0.33
3-Phenylpropyl acetate	15.74	5.94
1-Methoxy-4-decene	15.98	0.18
Cinnamyl acetate	16.98	0.23
6,10-Dimethyl-5,9-undecadien-2-one	17.02	0.35
Hexadecane	19.19	0.05

Four of the main aroma components of *M. variegatus* (Vittad.) Tul. & C. Tul., i.e., in decreasing importance order, isobutyl acetate (fruity and floral), ethyl 2-methylpropanoate (fruity), 2-methypropyl 2-methyl-2-butenoate (green, fruity, sweet), and 2-methylpropyl propanoate (fruity odor), explain the intense fruity component of its complex sweet aroma which is completed by a "cherry laurel" nuance, probably due to 2-methylpropanol (camphor odor) and by a "lightly spiced" segment that could derive from 3-phenylpropyl acetate (balsamic, spicy).

Octavianina Asterosperma

In *O. asterosperma* (Figure 13), dimethyl sulphide, methyl 2-propenoate, methyl 2-methylpropanoate, ethyl propanoate, methyl butanoate, methyl 2-methylbutanoate, ethyl methacrylate, methyl 2-methyl-3-oxobutanoate, and 3-octanone were found (Table 19).

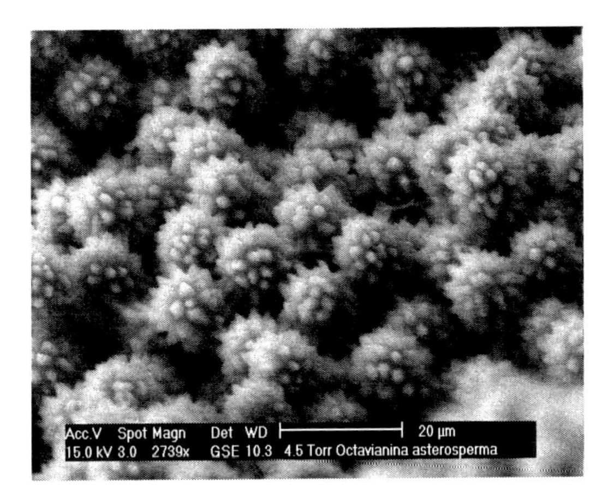

Figure 13. Basidiospore di *Octavianina asterosperma*.

Table 19. VOCs identified in *Octavianina asterosperma* and corresponding per cent area

Compound	Retention time [min.]	Area %
Dimethyl sulfide	1.70	7.32
Methyl 2-propenoate	2.15	2.89
Methyl 2-methylpropanoate	2.80	18.86
Ethyl propanoate	3.17	3.39
Methyl butanoate	3.32	2.69
Methyl 2-methylbutanoate	4.23	12.39
Ethyl methacrylate	4.39	1.45
Ethyl 2-butenoate	5.58	0.35
Methyl 2-methyl-3-oxobutanoate	7.17	1.17
3-Octanone	8.51	6.39
Cis-α-Bisabolene	17.83	0.05

The main components of *O. asterosperma* VOCs, i.e., ethyl 2-methylpropanoate (18.86%) and methyl 2-methylbutanoate (12.39%), both characterised by a fruity aroma, match definition of its aroma as "fruit-scented or candy-like" (Montecchi & Sarasini, 2000). Dimethyl sulphide (7.32%) perhaps along with 3-octanone (fournished of a cheesy odour), completes the above definition.

CONCLUSION

In conclusion, this study showed that the VOCs found in the different species of truffles and false-truffles allowed a better characterization of their respective aromas. Some of the analyzed species contained dimethyl sulphide as the main scent component, allowing to explain their particular aromas; other species showed relevant amount of unsaturated and aromatic compounds, that could give their sporophores various pleasant odours; sporophores of some other species, characterized by the presence of aliphatic esters, had a fruity aroma, and finally, others species are characterized by the presence of relevant amounts of mono and sesquiterpenes.

So far VOCs of truffle and false truffle species and varieties subjected to SPME-GC-MS analyses in this investigation were unknown. This gap of knowledge rendered difficult to exaustively indicate their sometimes complex aromas which remained unclear or undefinite.

REFERENCES

Angeletti, M; Landucci, A; Contini, M; Bertuccioli, M. Caratterizzazione dell'aroma del tartufo mediante l'analisi gas cromatografia dello spazio di testa. *In Atti II Congresso Internazionale sul Tartufo*, Spoleto, Italy, 1988, 505-509.

Balestreri, F; Martini, D; Vietti, M. Determinazione mediante HPLC dei componenti dell'aroma del Tuber melanosporum. *In Atti XII Congresso Nazionale di Merceologia*, 1986, Torino, Italy.

Balestrieri, F; Martini, D; Vietti, M. Sulla variazione dell'aroma del Tuber magnatum durante il processo di maturazione. *Industrie Alimentari*, 1988, 27, 423.

Bellesia, F; Pinetti, A; Bianchi, A; Tirillini. B. I composti solforati dell'aroma del tartufo: loro evoluzione durante la conservazione. *Atti Soc. Nat. Mat. Modena*, 1996a, 127, 177-187.

Bellesia, F; Pinetti, A; Bianchi, A; Tirillini, B. The volatile organic compounds of white truffle (Tuber magnatum Pico) from middle Italy. *Flavour Fragr. J.*, 1996b, 11, 239-243.

Bellesia, F; Pinetti, A; Bianchi, A; Tirillini, B. The volatile organic compounds of Tuber uncinatum from middle Italy. *J. Essent. Oil Res.*, 1998a, 10, 483-488.

Bellesia, F; Pinetti, A; Bianchi, A; Tirillini, B. The volatile organic compound of black truffle (Tuber melanosporum Vitt.) from middle Italy. *Flavour Fragr. J.*, 1998b, 13, 56-58.

Bellesia, F; Pinetti, A; Tirillini, B; Bianchi, A. Temperature-dependent evolution of volatile organic compounds in Tuber borchii from Italy. *Flavour Fragr. J.*, 2001, 16, 1-6.

Bellina Agostinone, C; D'Antonio, M; Pacioni, G. Odour composition of the summer truffle, Tuber aestivum. *Trans. Br. Mycol. Soc.*, 1987, 884, 568-569.

Boyd-Boland, A; Chai, M; Luo, Y; Zang, Z; Yang, M; Pawliszyn, J; Gorecki, T. Solvent free sample preparation techniques based on fiber and polymer technologies. *Environ. Sci. Technol.*, 1994, 28, 569A-574A.

Calvo-Gomez, O; Morales-Lopez, J; Lopez, MG. Solid-phase microextraction-gas chromatographic-mass spectrometric analysis of garlic oil obtained by hydrodistillation, *J. Chromat.*, A2004, 1036, 91-93.

Cerone, G; Rana, GL; Tagliavini, O. Carta delle vocazioni tartufigene della *Basilicata. Boll. Uff. Reg.*, Basilicata2004, 21 (Suppl. ord.), Parte I, 7-54 ().

Chin, H; Bernhard, R; Rosemberg, M. Solid phase microextraction for cheese volatile compounds analysis. *J. Food Sci.*, 1996, 61, 1118-1129.

Clark, J; Bunch, J. Qualitative and quantitative analysis of flavor additives on tobacco products using SPME-GC-MS. *J. Agr. Food Chem.*, 1997, 45, 844-849.

Claus, R; Hoppe, HO; Karg, H. The secret of truffles: a steroidal pheromone? *Experentia*, 1981, 37, 1178-1179.

Elmore, S; Erbahdir, M; Mottram, D. Comparison of dynamic headspace concentration on tenax with SPME for analysis of aroma volatiles. *J. Agr. Food Chem.*, 1997, 45, 2638-2641.

Estrada-Torres, AT; Gaither, W; Miller, DL; Lado, C; Keller, HW. The mixomycete genus Schenella: mophological and DNA sequence evidence for synonymy with the gasteromycete genus Pyrenogaster. *Mycologia*, 2005, 97, 139-149.

Fiecchi, A; Galli Kienle, M; Scala, A; Gabella, P. Bismethylthiomethane, an odorous substance from white truffle. *Tetrahedron Lett.*, 1967, 18, 1681-1682.

Fiecchi, A. Odour composition of truffles. *In Atti II Congresso Internazionale sul Tartufo*, Spoleto, Italy, 1988, 497-500.

Flament, I; Chevallier, C; Dobonneville, C. Analysis of volatile flavor constituents of Perigord black truffle (Tuber melanosporum Vitt.). *Riv. Ital. EPPOS*, 1990, 9, 280-299.

Hanssen, HP; Kühne, B. Aroma compounds in canned truffles. *In Atti II Congresso Internazionale sul Tartufo*, Spoleto, Italy, 1988, 501-504.

Hibbet, DS; Binder, M; Bischoff, JF; Blackwell, M; Cannon, PF; Eriksson, OE; Huhndorf, S; James, T; Kirk, PM; Mclaughlin, DJ; Powell, MJ; Redhead, S; Schoch, CL; Spatafora, JW; Stalpers, JA; Vilgalys, R; Aime, MC; Aptroot, A; Bauer, R; Begerow, D; Benny, GL; Castelbury, LA; Crous, PW; Dai, YC; Gams, W; Geiser, DM; Griffit, GW; Gueidan, C; Hawksworth, DL; Hestmark, G; Hosaka, K; Humber, RA; Hyde, KD; Ironside, JE; Koljag, U; Kurtzman, CP; Larsson, KH; Lichtward, R; Longcore, J; Miadlikowska, J; Miller, A; Moncalvo, JM; Mozley-Standridge, S; Oberwinkler, F; Parmasto, E; Reeb, V; Rogers, JD; Roux, C; Ryvarden, L; Sanpaio, JP; Schubler, A; Sugiyama, J; Thorn, RG; Tibell, L; Untereiner, WA; Walker, C; Wang, Z; Weir, A; Weiss, M; White, MM; Winka, K; Yao, YJ; Zhang. N. A higher-level phylogenetic classification of the fungi. *Mycol. Res.*, 2007, 111, 509-547.

Kim, JW; Kim, YS; Kyung, KH. Inhibitory activity of essential oils of garlic and onion against bacteria and yeasts, *J. Food Prot.*, 2004, 67, 499-504.

Komatsu M; Uchiyama T; Omura S; Cane DE; Ikeda H. Genome-minimized Streptomyces host for the heterologous expression of secondary metabolism. *Proc. Natl Acad. Sci.*, 2010, 107, 2646-2651.

Malençon, G; Riousset, L; Pyrenogaster pityophilus, G; Malençon et, L. Riousset, noveau genre et nouvelle espèce de Gastéromycète (Geastraceae), *Bull. Soc. Mycol. France*, 1977, 113, 173-187.

Matich, A; Rowan, R; Banks, N. Solid-phase microextraction for quantitative headspace sampling of apple volatiles. *Anal. Chem.*, 1996, 68, 4114-4118.

Menotta, M; Gioacchini, AM; Amicucci, A; Buffalini, M; Sisti, D; Stocchi, V. Headspace solid-phase microextraction with gas chromatography and mass spectrometry in the investigation of volatile organic compounds in an ectomycorrhizae synthesis system. Rapid Commun. *Mass Spectrom.*, 2004, 18, 206-210.

Montecchi, A; Borelli, M. Funghi ipogei raccolti nelle località vittadiniane, *Rivista di Micologia*, 1990, 33, 278-280.

Montecchi, O; Sarasini, M. Funghi ipogei d'Europa. A.M.B. Ed., Trento, Italy, 2000, 714 pp.

Ney, KH; Frietag, WG. Truffle aroma. *Gordian*, 1980, 80, 214-216.

Pacioni, G; Bellina Agostinone, C; D'Antonio, M. Odour composition of Tuber melanosporum complex. *Mycol. Res.*, 1990, 94, 201-204.

Pelusio, F; Nilsson, T; Montanarella, L; Tilio, R; Larsen, B; Facchetti, S; Madsen, JØ. Heaspace solid-phase microextraction analysis of volatil organic sulfur compounds in black and white truffle aromas. *J. Agr. Food Chem.*, 1995, 43, 2138-2143.

Poumarat, S; Neville, P. Une espèce rare et curieuse de "Gasterpmycetes": Pyrenogaster pityophilus Malençon & Riousset, Bull. Fédér. *Assoc. Mycol. Médit.*, 1997, 11, 14-21.

Rana, GL; Marino, R; Camele, I; Altieri, L. Nuove acquisizioni sui funghi ipogei della Basilicata. *Micologia Italiana*, 2008, 37, 52-64.

Rana, GL; Signore, SF; Fascetti, S; Marino, R; Mang, SM; Zotta, T. Seconda segnalazione del Pachyphloeus prieguensis in Italia ed acquisizioni recenti sui funghi ipogei lucani e pugliesi. *Micologia e Vegetazione Mediterranea*, 2010, 25, 47-80.

Shaath, NA; Flores, FB. Egyptian onion oil, *Dev. Food Sci.*, 1998, 40, 443-453.

Signore, SF; Rana, GL; Lolli, G; Laurita, A. Schenella pityophila, un raro gasteromicete rinvenuto nel Salento, *Micol. Veget. Medit.*, 2008, 23, 135-145.

Song, J; Gardner, BD; Holland, JF; Beaudry, RM. Rapid analysis of volatile flavor compounds in apple fruit using SPME and GC/time-of-flight mass spectrometry. *J. Agric. Food Chem.*, 1997, 45, 1801-1807.

Steffen, A; Pawliszyn, J. Analysis of flavor volatiles using headspace solid phase microextraction. *J. Agr. Food Chem.*, 1997, 44, 2187-2191.

Talou, T; Delmas, M; Gaset, A. Principal constituents of black truffles (Tuber melanosporum) aroma. *J. Agr. Food Chem.*, 1987, 35, 774-777.

Talou, T; Delmas, M; Gaset, A. Advances in the detection of black truffles. *Riv. Ital. EPPOS*, 1990, 9, 300-316.

Tirillini, B; Verdelli, G; Paolocci, F; Ciccioli, P; Frattoni, M. The volatile organic compounds from mycelium of Tuber borchii Vitt. *Phytochemistry*, 2000, 55, 983-985.

Venturella, G; Saitta, A; Morara, M; Zambonelli, A. Pyrenogaster pityophilus (Geastraceae), a new record from Sicily (S-Italy), *Fl. Medit.*, 2004, 14, 263-266.

Vittadini, C. Monographia Tuberacearum. Rusconi: Milano, 1831, pp. 88.

Yang, X; Peppard, T. Solid-phase Microextraction for Flavor Analysis, *J. Agric. Food Chem.*, 1994, 42, 1925-1930.

Yu, TH; Wu, CM; Liou, YC. Volatile compounds from garlic, *J. Agric. Food Chem.*, 1989, 37, 725-730.

Zeppa, S; Gioacchini, AM; Guidi, C; Guescini, M; Pierleoni, R; Zambonelli, A; Stocchi, V. Determination of specific volatile organic compounds synthesised during Tuber borchii fruit body development by solid-phase microextraction and gas chromatography/mass spectrometry. Rapid Commun. *Mass Spectrom.*, 2004, 18, 199-205.

Biographical Sketch

Maurizio D'auria

Prof. Maurizio D'Auria was born at Roma in 1953 and was graduated in Chemistry in 1977.

From 1979 until 1981 he had a fellowship of Fondazione Donegani at the National Academy of Lincei.

From 1981 until 1992 he was researcher at the Chemistry Department of the University of Rome "La Sapienza".

From 1986 until 1991 he had a cooperation agreement with Centro di Studio per la Chimica delle Sostanze Organiche Naturali of CNR.

In 1988 he was visiting professor at the Silesian Technical University of Gliwice in Poland.

From 1992 until 2001 he was associate professor at the Science Faculty of the University of Basilicata.

From 2001 he is full professor at the Agricultural Faculty of the University of Basilicata and then at the Department of Science of the same University. His H-index is 30.

From 1994 until 1996 è stato Presidente della Sezione Basilicata della Società Chimica Italiana. From 2000 he is member of the board of directors of CINMPIS (Consorzio Interuniversitario Nazionale Metodologie e Processi Innovativi di Sintesi).

From 2000 until 2014 he was coordinator of the International Doctorate in Chemical Sciences.

During his career he was teacher of several courses in organic chemistry (Organic Chemistry, Chemistry of Natural Organic Compounds, Organic Chemistry for Synthesis, Advanced Organic Chemistry, Chemistry of Heterocyclic Compounds). in a.a 2012-2013 he taught "Petroleum chemistry" in the 2nd Level Master on Petroleum Geosciences organized by the University of Napoli "Federico II" and the University of Basilicata. In a.a. 2012-2013 we maintained the same course in the 2nd Level Master on Petroleum Geosciences organized by the University of Basilicata. In the a.a. 2014-2015 he teaches Petroleum Chemistry for the Second Cycle Programme (Master Degree) in Geosciences and Georesources.

Publications of the last three years

1. **D'Auria, M.; Racioppi, R.; Rana, G. L.** Volatile organic compounds of *Schenella pityophilus*. *Nat. Prod. Res.* **2013**, *27*, 41-44.
2. **Todaro, L.; Dichicco, P.; Moretti, N.; D'Auria, M.** Effect of combined steam and heat treatments on extractives and lignin in sapwood and heartwood of Turkey oak (*Quercus cerris* L.) wood. *Bioresources* **2013**, *8*, 1718-1730.
3. **D'Annibale, A.; D'Auria, M.; Prati, F.; Romagnoli, C.; Stoia, S.; Racioppi, R.; Viggiani, L.** Paternò-Büchi reaction versus hydrogen abstraction in the photochemical reactivity of alkenyl boronates with benzophenone. *Tetrahedron* **2013**, *69*, 3782-3795.
4. **D'Auria, M.** A new proposal for the estimation of the aromatic character. *Lett. Org. Chem.* **2013**, *10*, 277-282.
5. **D'Auria, M.; Guarnaccio, A.; Racioppi, R.; Santagata, A.; Teghil, R.** Synthetic approach to and characterization of a fullerene-DTBT-fullerene triad. *Synlett* **2013**, *24*, 943-946.
6. **Attanasi, O. A.; Bianchi, L.; D'Auria, M.; Mantellini, F.; Racioppi, R.** Novel tetrahydropyridazines by unusual aza-Diels-Alder reaction of electron-poor 1,2-diaza-1,3-dienes with electron-poor alkenes under solvent free conditions. *Curr. Org. Synth.* **2013**, *10*, 631-639.
7. **D'Auria, M.** On a dispute between Ciamician and Paternò. *EPA Newsletter* **2013**, *84*, 106-109.
8. **D'Auria, M.; Racioppi, R.** Oxetane synthesis through the Paternò-Büchi reaction. *Molecules* **2013**, *18*, 11384-11428.

9. **D'Auria, M.** At the origin of photochemistry. The photochemical behaviour of santonin. Some documents in this field. *EPA Newsletter*, **2013**, *85*, 79-86.
10. **Attanasi, O. A.; Bianchi, L.; D'Auria, M.; Favi, G.; Mantellini, F.; Racioppi, R.** Aza-Diels-Alder reaction between 1,2-diaza-1,3-dienes and β-aryl-α, β-unsaturated carbonyl compounds. Easy one-pot entry to 2'-oxoimidazo[1',5'-*f*]tetrahydropyridazine. *Curr. Org. Synth.* **2013**, *10*, 951-960.
11. **D'Auria, M.; Masini, N.** L'opera scientifica di Francesco Mauro. *Rendiconti dell'Accademia Nazionale delle Scienze detta dei XL, Memorie di Scienze Fisiche e Naturali, Serie V,* **2013**, *37, p. II, t. II*, 59-70.
12. **D'Auria, M.** A new index for the estimation of the aromatic character - II. *Lett. Org. Chem.* **2014**, *11*, 250-258
13. **D'Auria, M.** The Paternò-Büchi reaction. In *Comprehensive Organic Synthesis, 2nd Edition*, Molander, G. A.; Knochel, P. (Eds.), Vol. 5, Elsevier, Oxford, 2014, pp. 159-199.
14. **Altieri, G.; Di Renzo, G. C.; Genovese, F.; Tauriello, A.; D'Auria, M.; Racioppi, R.; Viggiani, L.** Olive oil quality improvement using a natural sedimentation plant at industrial scale. *Biosystems Engineering* **2014**, *122*, 99-114.
15. **Acquaviva, V.; D'Auria, M.; Racioppi, R.** Changes in aliphatic ester composition in white wines during exposition to light. An HS-SPME-GC-MS study. *J. Wine Res.* **2014**, *25*, 63-74.
16. **D'Auria, M.** Francesco Mauro, scienziato. In *Francesco Mauro - un chimico lucano*, D'Auria, M.; Colella, C.; Masini, N. (Eds), Edizioni Scientifiche Italiane, Napoli, 2014, pp 123-135 ISBN 978-88-495-2778-0
17. **D'Auria, M.** La chimica del molibdeno oggi - i poliossimetallati. In *Francesco Mauro - un chimico lucano*, D'Auria, M.; Colella, C.; Masini, N. (Eds), Edizioni Scientifiche Italiane, Napoli, 2014, pp 137-148 ISBN 978-88-495-2778-0
18. **D'Auria, M.** At the origin of photochemistry. The photochemical behavior of santonin. Some considerations on the published documents. *EPA Newsletter* **2014**, *86*, 76-80
19. **D'Auria, M.** Italian photochemistry meeting 2013. *EPA Newsletter* **2014**, *86*, 102-106.
20. **D'Auria, M.** A new index for the estimation of the aromatic character - IV. *Lett. Org. Chem.* **2014**, *11*, 657-663.

21. **D'Auria, M.; Racioppi, R.; Rana, G. L.; Laurita, A.** Studies on volatile organic compounds of some truffles and false truffles. *Nat. Prod. Res.* **2014**, *28*, 1709-1717.
22. **D'Auria, M.** A new index for the estimation of the aromatic character - III. *Lett. Org. Chem.* **2014**, *11*, 731-735.
23. **D'Auria, M.** A DFT study of the photochemical dimerization of methyl 3-(2-furyl) acrylate and allyl urocanate. *Molecules* **2014**, *19*, 20482-20497.
24. **D'Auria, M.** At the origin of photochemistry. An article of Paternò written in 1875. *EPA Newsletter* **2014**, *87*, 96-99.
25. **Todaro, L.; D'Auria, M.; Langerame, F.; Salvi, A. M.; Scopa, A.** Surface characterization of untreated and hydro-thermally pre-treated Turkey oak woods after UV-C irradiation. *Surf. Interface Anal.* **2015**, *47*, 206-215.
26. **Todaro, L.; Rita, A.; Cetera, P.; D'Auria, M.** Thermal treatment modifies the calorific value and ash content in some wood species. *Fuel*, **2015**, *140*, 1-3.
27. **D'Auria, M.; Guarnaccio, A.; Racioppi, R.; Santagata, A.; Teghil, R.** Synthesis ad photophysical properties of some dithienylbenzo[*c*]thiophene derivatives. *Heterocycles*, **2015**, *91*, 313-331.
28. **D'Auria, M.** A new index for the estimation of the aromatic character - V. *Lett. Org. Chem.* **2015**, *12*, 233-236.
29. **Mang, S. M.; Racioppi, R.; Camele, I.; Rana, G. L.; D'Auria, M.** Use of volatile metabolite profiles to distinguish three *Monilinia* species. *J. Plant Pathol.* **2015**, *97*, 55-59.
30. **D'Auria, M.** A new index for the estimation of the aromatic character - VII. *Lett. Org. Chem.* **2015**, *12*, 402-406
31. **D'Auria, M.** The presence of photochemistry in the literature. A personal selection between the articles published in 2015. *EPA Newsletter* **2015**, *88*, 56-84
32. **D'Auria, M.** At the origin of photochemistry. The dimerization of cynnamic acid derivatives. *EPA Newsletter* **2015**, *88*, 150-156.
33. **D'Auria, M.** A new index for the estimation of the aromatic character - VI. *Lett. Org. Chem.* **2015**, *12*, 482-490.
34. **D'Auria, M.** A new index for the estimation of the aromatic character - VIII. *Lett. Org. Chem.* **2015**, *12*, 549-559.
35. **D'Auria, M.; Racioppi, R.** The effect of drying on the composition of volatile organic compounds in *Rosmarimuns officinalis, Laurus*

nobilis, Salvia officinalis and *Thymus serpyllum*. A HS-SPME-GC-MS study. *J. Essent. Oil Bearing Plants* **2015**, *18*, 1209-1223.

36. **D'Auria, M.; Frenna, V.; Monari, M.; Palumbo-Piccionello, A.; Racioppi, R.; Spinelli, D.; Viggiani, L.** $Ru(bpy)_2Cl_2$: a catalyst able to shift the course of the photorearrangement in the Boulton–Katritzky reaction. *Tetrahedron Lett.* **2015**, *56*, 6598-6601.
37. **D'Auria, M.** At the origin of photochemistry. Leone Maurizio Padoa. *EPA Newsletter*, **2015**, *89*, 74-78.
38. **D'Auria, M.** A new index for the estimation of the aromatic character - IX. *Lett. Org. Chem.* **2016**, *16*, 33-43.
39. **Guarnaccio, A.; D'Auria, M.; Racioppi, R.; Mattioli, G.; Amore Bonapasta, A.; De Bonis, A.; Teghil, R.; Prince, K. C.; Acres, R. G.; Santagata, A.** Thiophene-Based Oligomers Interacting with Silver Surfaces and the Role of a Condensed Benzene Ring. *J. Phys. Chem. C* **2016**, *120*, 252–264.

In: Volatile Organic Compounds
Editor: Julian Patrick Moore

ISBN: 978-1-63485-370-5

Chapter 7

REMOVAL OF VOLATILE ORGANIC COMPOUNDS IN AIR BY TOTAL CATALYTIC OXIDATION PROMOTED BY CATALYSTS

Rosana Balzer
Department of Engineering and Exact Sciences,
Federal University of Parana, UFPR, Brazil

ABSTRACT

Volatile organic compounds (VOCs) are hazardous highly toxic pollutants that cause a number of environmental and human health problems. They are released during a wide range of industrial, transportation and commercial activities and their emissions have reached high levels. There is therefore a need for the development of techniques which are both economically feasible and able to effectively remove these pollutants. Several VOCs removal techniques, including physical, chemical and biological methods, have been described in the literature. Catalytic oxidation has been acknowledged as the most effective approach, mainly due to its high degradation efficiency, low energy cost, the potential for the removal of low concentrations of VOCs and the low thermal NO_x emissions involved. This capther was to investigate the catalytic behavior of catalysts with low production costs, involving a simple method of preparation without the use of reagents which are harmful to the environment, which are active at low temperature, have high redox potential for catalytic oxidation reactions and do not result in the formation of byproducts. In this regard, catalysts

Co_{10}/γ-Al_2O_3-CeO_2 and Co_{20}/γ-Al_2O_3-CeO_2; catalysts $(SiO_{2(1\text{-}x)}Cux)$ were developed which exhibit high catalytic activity for converting BTX compounds and WO_3-based catalyst to investigate its activity catalytic in the total oxidation of the volatile organic compounds known as BTX. For a range of low temperatures the only reaction products were CO_2 and H_2O.

1. INTRODUCTION

Volatile organic compounds (VOCs) are considered to be a major contributor to air pollution, since they are harmful to the environment and human health. Catalytic oxidation has been recognized as the most effective method to reduce air pollution and the concentration of VOCs emitted to the atmosphere. VOCs are dangerous pollutants with high toxicity, which are present in waste gases generated by motor vehicles and, more importantly, by industrial processes that use solvents. The main advantages of catalytic oxidation is are that it can occur at relatively low temperatures and there is no formation of by-products. The desired reaction products are carbon dioxide (CO_2) and water (H_2O) [1-9].

The contamination of atmospheric air, soil and groundwater by volatile organic compounds has been of concern in recent years mainly due to the hight levels of petroleum hydrocarbon compounds emitted into the atmosphere with harmful consequences. The volatile organic compounds belong to a wide range of toxic air pollutants which include hydrocarbons, aromatics and molecules containing oxygen, nitrogen, sulfur and halogen. Volatile organic compounds present in air are derived from various industrial processes and they are harmful to human health and to the environment even at low concentrations. It is widely recognized that the emission of BTX compounds is a critical environmental problem and thus several techniques for the reduction of these compounds have been investigated [1, 10-14].

The potential pollutant gasoline is directly related to aromatic hydrocarbons, namely benzene, toluene and xylenes (known as BTXs compound or simply BTXs). Due to the high toxicity of BTX compounds and the resulting adverse effects in contaminated systems, the search for control and remediation procedures is of great importance [6-8, 15, 16].

Research on the development of new technologies that enable the replacement of polluting energy sources with clean renewable sources has stimulated an increase in research studies on heterogeneous catalysis. In this context, the emission of BTX compounds, which are highly toxic pollutants associated with industrial activity, causes numerous environmental problems. Catalytic oxidation does not require the addition of fuel, which reduces the energy consumption and avoids the formation of thermal NOx. This process is regarded as effective because it operates at low temperatures and there is no formation of byproducts from thermal oxidation [3, 13, 14, 17, 18].

Catalytic oxidation has been recognized as an efficient method to reduce air pollution, particularly atmospheric concentrations of volatile organic compounds. The advantages of catalytic oxidation include the fact that it can operate at relatively low temperatures and no byproducts are formed [11, 16, 19-21].

The catalytic oxidation of BTX to control gaseous industrial emissions is one of the most promising environmental technologies. Catalytic oxidation is one of the most important methods for the removal of BTX from the air many environmental problems, such as toxicity and petrochemical smog are related to BTX emissions [22, 23].

The catalysts commonly used for the catalytic oxidation are based on noble metals, which often perform better when compared with non-noble metal based catalysts. As an example, platinum catalysts (Pt) and palladium (Pd) that are typically applied to promote these reactions. However, due to the high cost of these metals, which can represent an economic obstacle to be employed in this process, these metals are being replaced by catalysts using non-noble metals. Therefore, search is increasingly economically viable catalysts and consisting of transition metals such as cobalt (Co), cerium (Ce), copper (Cu), manganese (Mn) and chromium (Cr) supported on matrices with high surface area as alumina (Al_2O_3) and silica (SiO_2), which have shown a high oxidation potential. It has been reported in the literature that CeO_2 has the potential to increase the degree of oxidation in reactions, due to the creation of active oxygen. The catalytic activity of CeO_2 has also been attributed to its ability to store oxygen. The incorporation of CeO_2 thus improves the performance of the redox catalyst, besides acting as a stabilizer of O_2 on the surface [2-5, 7, 10, 11].

The observed differences in catalytic activity on oxidation reactions displayed by the catalysts have been attributed to a combination of factors, including their physical and chemical properties, which are related to the method used for its preparation. Therefore, the development of new catalysts that meet these conditions and that have superior performance on the above process becomes a challenge to researchers [4, 5, 7, 10, 11].

The objective of this capther was to investigate the catalytic behavior of catalysts with low production costs, involving a simple method of preparation without the use of reagents which are harmful to the environment, which are active at low temperature, have high redox potential for catalytic oxidation reactions and do not result in the formation of byproducts.

2. Catalytic Oxidation of Volatile Organic Compounds

The oxidation of an organic compound may be defined as the addition of oxygen, hydrogen removal or the removal of electrons of this compound.

Although the oxidation reactions of organic compounds by oxygen are spontaneous, the catalytic oxidation reactions are usually promoted by metallic species, or require the use of catalysts. The need to utilize a metal catalyst occurs due to high activation energy which is required for these reactions.

Therefore an oxidation reaction by air, although difficult to start, usually occurs to the formation of carbon dioxide (CO_2) and water (H_2O). For example, the total oxidation of n-hexane is shown in the following equation 1.

$$C_6H_{14(g)} + 9{,}5O_{2(g)} \rightarrow 6CO_{2(g)} + 7H_2O_{(g)} \qquad (1)$$

Catalyzed oxidation reactions by metals are classified into two types: homolytic and heterolytic. The homolytic employs catalysis soluble salts of transition metals (homogeneous reaction), such as acetates or metal oxides (heterogeneous reaction). In heterolytic catalysis, organic substrates are coordinated to the transition metal. [4, 5, 7, 10, 11]

Another important factor for oxidation reactions is the necessity of using metallic species which have more than one oxidation state. The current interest in the development of cleaner processes has diverted attention to the use of oxygen (O_2) as an oxidizing agent in combination with transition metal catalysts. In reactions in aqueous media have also been given great importance in the use of hydrogen peroxide (H_2O_2) as oxidant [4, 5, 7, 10]. In order to follow the environmental restrictions, the catalytic oxidation is playing an important role in projects of a more sustainable chemistry.

3. Mechanism Total Catalytic Oxidation Promoted by Catalysts

It is well known that the oxidation of BTX compounds promoted by solid oxide catalysts can be proceed by the Mars-van Krevelen mechanism, in which the key steps are the supply of oxygen by the reducible oxide, the introduction of the (originating from the oxide lattice) into the substrate molecule and the re-oxidation of the reduced solid by the oxygen-containing gaseous phase, which is the rate-determining step of the reaction. According to the Mars-van Krevelen mechanism, the BTX oxidation rate is determined by the concentration of BTX chemisorbed onto the metal oxide particles close to the periphery of the metal-support interface [5, 7, 14].

According Figure 1, the active oxygen species formed on the metal-support interface or on the support close to the metal directly participate in the catalytic oxidation, and the metal coated on the support serves as the adsorption site for BTXs.

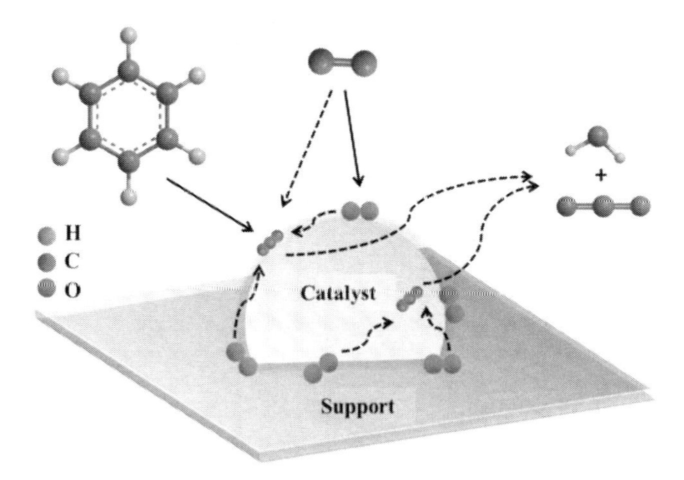

Figure 1. Catalytic oxidation of volatile organic compounds promoted by catalysts to CO_2 and H_2O [7].

The catalytic cycle involves the chemisorption of BTX onto Cu particles, migration of the chemisorbed BTX to the metal-support interface, O_2 activation on the defect sites in the support or at the metal-support interface, the formation of active oxygen species and the reaction between the chemisorbed BTX and active oxygen species at the interface [5-7, 9, 14].

4. Catalyst Performance for Volatile Organic Compounds Oxidation

In Figures 2 and 3 the conversion through the oxidation of n-hexane and n-heptane, respectively, can be observed, according to the reaction temperature.

The two catalysts showed similar profiles, that is, with an increase in temperature there was, an increase in the conversion of n-hexane and n-heptane for both the catalysts. As expected, with a higher metal content in the samples the conversion capacity was enhanced.

The highest conversion results for both n-hexane and n-heptane were obtained by the Co_{20}/γ-Al_2O_3-CeO_2 catalyst, that is, the sample which had a metal content of 20% cobalt. In the oxidation of n-hexane with the catalyst Co_{10}/γ-Al_2O_3-CeO_2 the conversion was 40-84% and for Co_{20}/γ-Al_2O_3-CeO_2 it reached 55-96%. The catalyst Co_{20}/γ-Al_2O_3-CeO_2 showed higher activity compared to Co_{10}/γ-Al_2O_3-CeO_2 [5, 24].

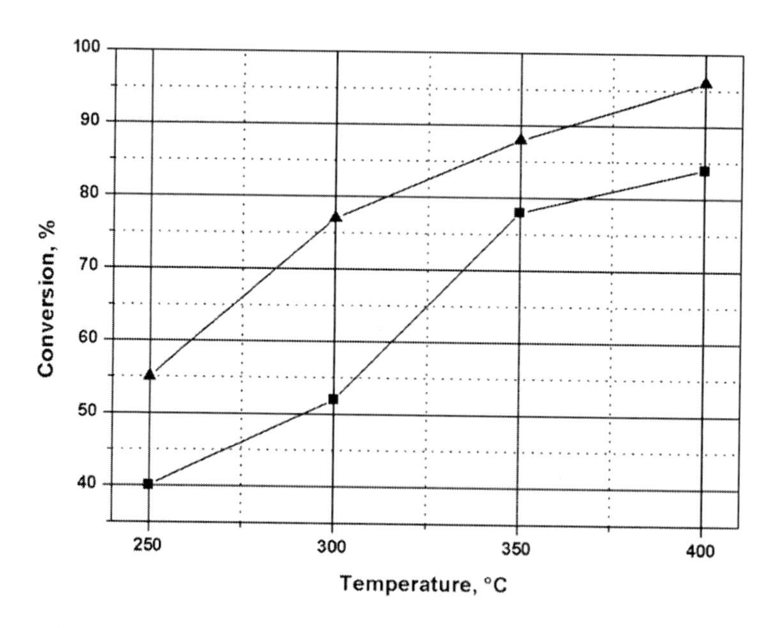

Figure 2. Conversion of n-hexane as a function of temperature: (■)Co_{10}/γ-Al_2O_3-CeO_2 and (▲)Co_{20}/γ-Al_2O_3-CeO_2.

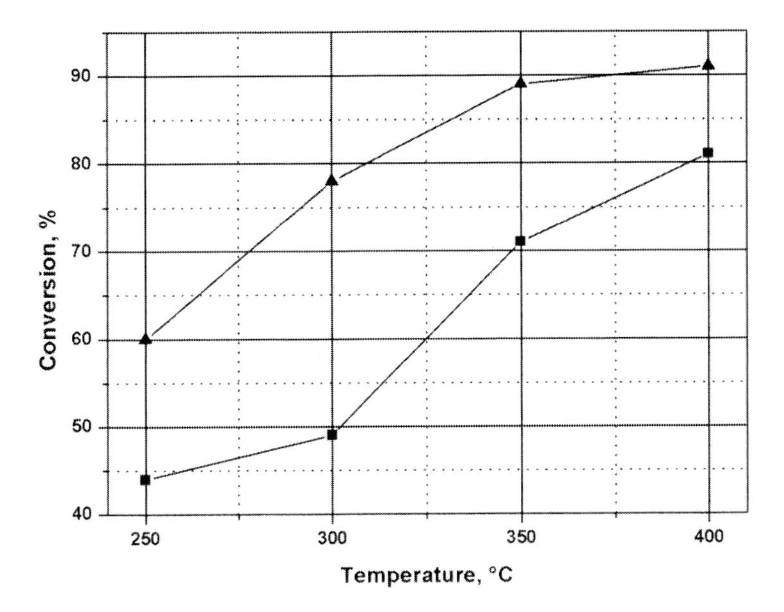

Figure 3. Conversion of n-heptane as a function of temperature: (■) $Co_{10}/\gamma\text{-}Al_2O_3\text{-}CeO_2$ and (▲) $Co_{20}/\gamma\text{-}Al_2O_3\text{-}CeO_2$.

The behaviors observed for n-heptane and n-hexane were similar, that is, the catalyst with the highest catalytic activity was $Co_{20}/\gamma\text{-}Al_2O_3\text{-}CeO_2$ and with an increase in temperature the rates of conversion also increased. As observed for the catalyst $Co_{10}/\gamma\text{-}Al_2O_3\text{-}CeO_2$, conversions were in the range of 44 to 81% and for $Co_{20}/\gamma\text{-}Al_2O_3\text{-}CeO_2$ they were in the range of 60 to 91%. The values for the activation energy of the reactions calculated using the Arrhenius equation were 15.9 and 10.8 kJ for the catalyst containing 10% cobalt and 13.3 and 9.1 kJ for that containing 20% cobalt, for the n-hexane and n-heptane oxidation reactions, respectively.

The surface properties of these catalysts are crucial to examining their catalytic activities. One factor to consider is the mobility of the oxygen atoms present, which are capable of carrying out the oxidation of the hydrocarbon under study. It has been shown that reduction of cerium oxide ($Ce^{4+} \rightarrow Ce^{3+}$) is not due to a direct release of O_2 into the gas phase, but rather to the interaction which occurs between the surface of the catalyst and the hydrocarbon. These reactions are driven by a greater capacity for the spontaneous release of oxygen from the Co_3O_4/CeO_2 system, even in the absence of a reducing agent. It is well known that the oxidation of hydrocarbons promoted by solid oxide catalysts can proceed via the Mars and van Kreveler mechanism [5, 24, 25].

The determination of reducible species at the surface of the catalyst and the temperature at which these species are reduced give important information regarding catalytic performance. The steps involved in Co_3O_4 reduction is still a controversial subject. There are two types of TPR spectra for Co_3O_4 reported in the literature: with one broad peak due to the single step for the Co_3O_4 reduction and with two peaks ascribed to a two-step process ($Co_3O_4 \rightarrow CoO \rightarrow Co$) [5, 25, 26].

The TPR spectrum for CeO_2 may contain one to three peaks. It is generally accepted that CeO_2 reduction at the surface occurs via a stepwise mechanism, starting with the reduction of the outermost layer of Ce^{4+} produced at lower temperatures (a peak at between 400-550°C), then a second peak at approximately 580-650°C, which is due to the formation of non-stoichiometric oxides of composition CeO_x and finally the reduction of the inner Ce^{4+} (CeO_2 to Ce_2O_3 bulk reduction) at temperatures above 750°C [26, 27].

The TPR profile of the $Co_{10}/\gamma\text{-}Al_2O_3\text{-}CeO_2$ sample shows a broad major H_2 consumption peak in the temperature range of 500-800°C that can be ascribed to the reduction of CeO_2 as described above. However, this broad peak may interfere with that associated with the reduction of Co^{2+} which interacts with CeO_2 to form Co. On the other hand, the TPR profile of the $Co_{20}/\gamma\text{-}Al_2O_3\text{-}CeO_2$ sample shows three major H_2 consumption peaks. The peak at 375ºC can be attributed to the reduction of the independent Co_3O_4 phase that weakly interacts with CeO_2 directly to form Co, while the second peak, centered at 590°C, results from the reduction of Co^{2+} or Co^{3+} ions strongly bound in the ceria matrix. Nevertheless, this second peak could also comprise the reduction of non-stoichiometric species, as mentioned above. The highest temperature TPR peak can be ascribed to the reduction of bulk CeO_2[27, 28].

The oxygen storage capacity of cerium oxide is associated with a fast Ce^{4+}/Ce^{3+} redox process, resulting in more oxygen being available for the oxidation process. The oxygen migration on the catalyst surface is important in oxidation reactions, where the oxidation-reduction cycles determine the catalytic activity. Thus, the redox properties of the catalyst play a key role in the process and are an important factor in determining the catalytic performance.

Another factor to be considered is the presence of active sites on the catalyst surface, such as Co^{2+} ions associated with partial oxidation of the hydrocarbon and Co^{3+} ions associated with total oxidation. The activity of

determined in the past (1-methoxy-3-methylbenzene, decane, anisole, 3-ethyl-5-methylphenol, 1,4-dimethoxybenzene, and 1,2,4-trimethoxybenzene) were detected, whereas 1-propanol, 3-octanol, 3-methyl-1-butanol, 5-hexen-2-ol, 3-nonanol, benzylic alcohol, methylphenols, and 3-methylbutylamine were absent.

Table 8. VOCs identified in samples of *Tuber brumale* (*) and its form *moschatum* and corresponding per cent area

Compound	**Sample**						
	1 (*)	**2**	**3**	**4**	**5**	**6 (*)**	**7**
acetaldehyde		0.1					
ethanol		0.3		0.2	0.7	0.2	0.7
dimethylsulphide	15.3	3.4	0.6	1.2	3.0	9.0	6.7
2-methylpropanal	4.2	11.6	3.0	1.9	2.4	0.8	2.4
methylethyl formate	4.3	0.8	1.1	5.7	10.7	4.3	1.1
butanone	2.5	11.8	7.3	0.8	1.3	1.5	2.7
ethyl acetate	0.3	0.2	0.3	0.1	0.2	1.1	
2-methyl-1-propanol	0.8	1.9		0.8	3.3		1.3
2-methylbutanal	12.8	18.6	5.2	7.2	3.0	2.5	8.6
3-methylbutanal	0.7	2.2	0.4	0.5	0.3	0.4	0.4
1-methylpropyl formate	21.7	4.0	2.1	4.3	16.5	8.8	8.7
2-methyl-1-butanol	1.0	2.5		1.1	1.1		1.6
1-methylpropyl acetate		0.1					
toluene		0.1					
1,3-dimethylbenzene		0.1					
styrene		0.2					
anisole	0.1	0.8	0.3	0.2		0.3	0.3
3-octanone		0.1					
1-octen-3-ol		0.1					
1-methoxy-3-methylbenzene	2.7			0.4		0.2	2.6
decane	0.1	0.8					
1-methoxy-4-methylbenzene		0.1					
3-ethyl-5-methylphenol	0.2			0.2			0.3
limonene		0.04					
1-propynylbenzene		0.1					
1,4-dimethoxybenzene	8.8	7.4		31.1	5.1	6.2	17.5
benzo[b]thiophene		0.5					
1,2,4-trimethoxybenzene	1.3	3.5	4.5	9.0	0.5	5.6	0.3

Tuber Melanosporum

Although the samples of *T. melanosporum* (Figure 9) were only two, the results of the analyses were uniform. As showed in Table 9, little amounts of dimethysulphide were found. The main VOCs were 2-methylbutanal, 2-methylpropanol, and 2-methyl-1-butanol. The following six esters, never detected before, were also found: ethyl 2-methylbutanoate, 2-methylpropyl 2-methylbutanoate, 2-methylbutyl 2-methylbutanoate, 2-methylpropyl 2-methylpropanoate, and 3-methylbutyl 2-methylpropanoate. As it is well known, the organic esters are strictly related to flavour. Therefore, these above-mentioned compounds could play an important role in the definition of the aroma of this truffle species.

Tuber Oligospermum

The results of SPME-GC-MS analyses of *T. oligospermum* are collected in Table 10. Only propanone was detected in two samples. The other VOCs were present with the former in mixture with different organic substances.

Figure 9. *Tuber melanosporum.*

Table 9. VOCs identified in *T. melanosporum* and corresponding per cent area

Compound	Sample	
	1	2
acetaldehyde	2.0	1.1
ethanol	3.4	1.6
dimethyllsulphide	1.4	2.1
2-methylpropanal	3.2	4.7
butanone	1.3	1.3
2-methylpropanol	12.7	9.8
2-methylbutanal	9.6	18.4
3-methylbutanal	1.3	1.1
1-methylpropyl formate	0.3	0.8
3-methyl-1-butanol	0.9	
2-methyl-1-butanol	24.9	25.5
hexanal	0.7	0.7
ethyl 2-methylbutanoate	0.2	0.3
2-methylpropyl 2-methylpropanoate		0.3
anisole		0.1
2-methylpropyl 2-methylbutanoate	0.9	0.7
2-methylbutyl 2-methylpropanoate	0.2	0.5
2-methylbutyl 2-methylbutanoate	0.4	1.4

Table 10. VOCs identified in *T. oligospermum* and corresponding per cent area

Compound	Sample		
	1	2	3
acetaldehyde			0.6
propanone		0.9	2.6
2-methylbutanal	0.6		
3-methylbutanal	2.0		

Tuber Panniferum

Finally, Table 11 summarises the results we obtained from the analyses of *T. panniferum* (Figure 10). It is noteworthy that in *T. panniferum*, as in *T. magnatum*, the main constituent of the aroma is 2, 4-dithiopentane.

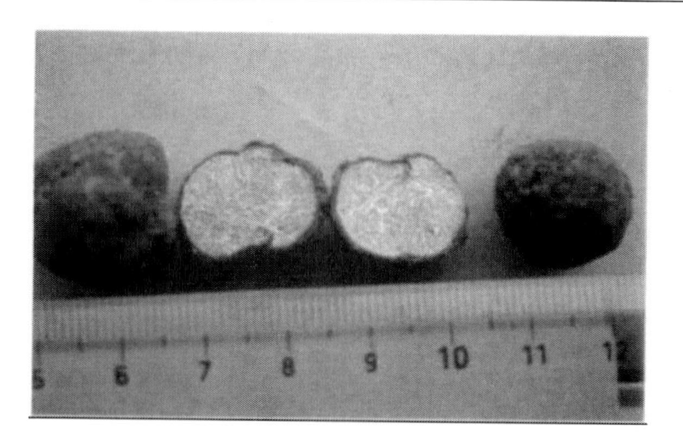

Figure 10. *Tuber panniferum.*

Table 11. VOCs identified in T. panniferum and corresponding per cent area

		1	**2**	**3**
9	2-methylpropanal	1, 0	1, 7	1, 8
12	butanone		16, 7	3, 8
17	2-methylbutanal		1, 6	1, 4
18	3-methylbutanal		1, 3	1, 4
23	dimethyldisulphide		0, 3	0, 5
33	2,4-dithiopentane	0, 5	18, 7	22, 3
39	3-octanone		3, 4	
40	benzaldehyde		1, 7	1, 5
72	6(*Z*),9(*E*)-heptadecadiene		0, 2	
73	1-heptadecene		0, 3	
74	1-nonadecene			0, 3

Other Ascomycetes

Tuber rufum Pico: Fr. var. *rufum*, *T. rufum* fo. *lucidum* (Bonnet) Montecchi & Lazzari and *Pachyphloeus conglomeratus* Berk. & Broomeshowed the presence of only dimethyl sulphide. In *Tuber dryophilum* Tul. & C. Tul., we detected the presence of some methoxy derivatives and some naphthalene compounds (Table 12). *C. meandriformis* Vittad. Broomeshowed the presence of only dimethyl sulphide.

cobalt oxide catalysts may be due to the presence of reducible oxides such as CeO_2. The oxygen required for the oxidation reaction to occur is thus provided by the second oxide, or cobalt oxide. The generation of active oxygen induced by cobalt active sites is therefore an important factor for catalytic activity observed. This shows that the formation of water can be facilitated by the ease of reduction of the cobalt species dispersed in the catalyst [28, 29].

The results show that the Co/γ-Al_2O_3-CeO_2 catalysts with levels of 10 and 20% of cobalt were effective in the oxidation reaction of hydrocarbons n-hexane and n-heptane, with high conversion even at temperatures as low 250ºC. The catalytic behaviors were mainly associated with the textural properties of the catalysts. The catalyst with the highest cobalt load presented the best performance, while for the catalyst with the lowest cobalt load the level of n-hexane and n-heptane conversion was the lowest over the whole temperature range. The physical-chemical characterizations indicated that the Co_{20}/γ-Al_2O_3-CeO_2 catalyst has a higher number of active sites available for the oxidation reactions. Only CO_2 and H_2O were observed as reaction products. The incorporation of CeO_2 provided a significant increase in the redox potential of these catalysts. The long-term tests carried out show excellent stability without significant loss of activity, which is related to the metal dispersion in the catalyst. Therefore, it can be concluded that the catalysts studied were efficient and stable in the catalytic oxidation of hydrocarbons to CO_2 and H_2O without the formation of byproducts [24, 29].

Figure 4 shows the conversion of BTXs as a function of the reaction temperature for the catalysts studied.

The catalyst $SiO_{2(0.9)}Cu_{0.1}$ showed higher catalytic activity toward the BTX compounds. The catalytic activity of the catalyst $SiO_{2(0.8)}Cu_{0.2}$ may be related to the excessive formation of agglomerated particles on the surface, reducing the metal dispersion, according to Graphical 1. As expected the catalyst $SiO_{2(0.97)}Cu_{0.03}$ had a lower catalytic activity than the other catalysts tested, due to its lower metal content. The values for the surface area, pore volume and pore diameter of the catalysts are presented in Graphical 1 [7].

As expected, there was a reduction in all of these parameters as the metal content increased. This reduction may be related to the high calcination of 550ºC (Graphical 1).The isotherms obtained were analyzed by comparison with the classification of the IUPAC (International Union of Pure and Applied Chemistry) and it was found that the catalysts have pores of regular cylindrical and/or polyhedral shape with open ends with isotermas type IV and histereses type H1 [7].

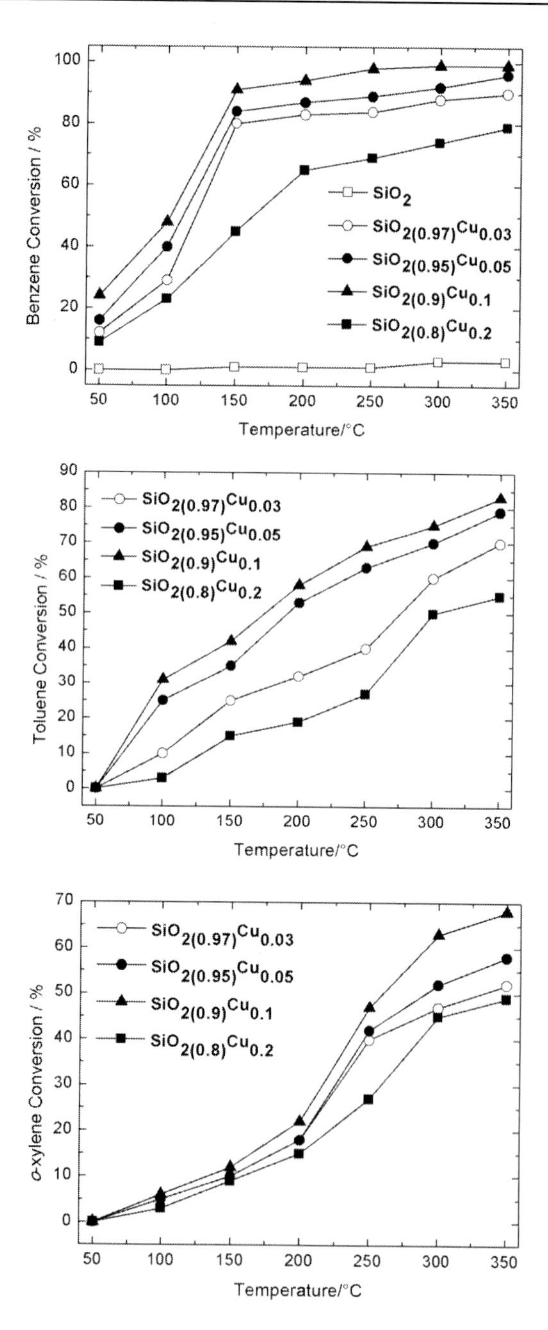

Figure 4. BTX conversion as a function of reaction temperature for different catalysts reaction under varying conditions.

Graphical 1. Textural data for the supported copper catalysts

Sample	BET ($m^2 g^{-1}$) before/after calcination	Vp ($cm^3 g^{-1}$) before/after calcination	Dp/nm before/after calcination	Average CuO size/nm[a]	Average Cu size/nm[b]	Cu loading wt%[c]	Cu dispersion%[d]
SiO_2	376/365	0.76/0.74	4.08/4.06	-	-	-	-
$SiO_{2(0.97)}Cu_{0.03}$	311/299	0.62/0.60	4.06/4.05	46	15.1	2.99	7.98
$SiO_{2(0.95)}Cu_{0.05}$	309/298	0.60/0.57	4.06/4.04	53	18.5	5.01	6.52
$SiO_{2(0.9)}Cu_{0.1}$	292/281	0.58/0.55	4.05/4.01	59	21.6	10.01	5.58
$SiO_{2(0.8)}Cu_{0.2}$	248/234	0.51/0.48	4.04/4.00	65	23.3	19.97	5.17

Vp = pore volume; Dp = pore diameter.

[a] Determined by XRD.

[b] Obtained by HRTEM.

[c] Calculated from semi-quantitative EDX data.

[d] Calculated using formula (Wu et al. 2011): ($D_{Cu} = 6n_s.M_{Cu}/\rho_{Cu}.N_A.d_{Cu}$), where n_s is the number of Cu atoms at the surface per unit area ($1.7x10^{19}$ m^{-2}); M_{Cu} is the molar mass of copper (63.546 g mol^{-1}); ρ_{Cu} is the density of copper (8.92 g cm^{-3}); N_A is Avogadro's number ($6.023x10^{23}$ mol^{-1}) and d_{Cu} is the average Cu particle size (obtained by HRTEM).

The X-ray diffractograms are where the diffraction peaks (2θ = 33, 35, 38, 48, 54, 58, 62, 65, 66 e 68°) are assigned to the reflections for the tenorita (CuO) phase, a monoclinic crystal system (JCPDS data file 045-0937). The peak at 2θ = 22° is characteristic of silicon oxide. The peaks are more intense for CuO (11-1) (2θ = 35°) and CuO (111) (2θ = 38). Therefore, it can be concluded that the large crystal size, good stability and close contact between the Cu and SiO_2 particles favors the catalytic performance of the Cu/SiO_2 catalyst.

On comparing the catalytic activity of the Cu/SiO_2 catalyst with other catalysts described in the literature which are highly active in BTX oxidation (Graphical 2), it can be seen that temperature $T_{benzene50}$ (benzene conversion up to 50%) is 106°C for $SiO_{2(0.9)}Cu_{0.1}$, 124°C for $SiO_{2(0.95)}Cu_{0.05}$ and 162°C for $SiO_{2(0.97)}Cu_{0.03}$. These values are comparable with those reported in literature for the catalyst Cu/γ-Al_2O_3, 1.5% Au/ZnO, 0.30% Pt/AC and 3% Au/V_2O_5/TiO_2, in fact, the benzene conversion exceeds 80% for the catalyst Cu/SiO_2 at 150°C, and is much higher than the values for the catalysts Cu/γ-Al_2O_3, 1.5% Au/ZnO, and 3% Au/V_2O_5/TiO_2 (0%, 80%, 63% respectively). Table 2 also shows the values for $T_{toluene50}$ (toluene conversion up to 50%) and $T_{o\text{-}xylene30}$ (o-xylene conversion up to 30%). The results of these experiments revealed that the SiO_2 support has no activity and reports in the literature for other catalysts indicate that interaction between Cu and SiO_2 could improve the catalytic activity of Cu/SiO_2 in relation to BTX oxidation [19, 24, 28, 29].

Graphical 2. Comparison of data published in the literature with those obtained in this study

Sample	BTX			BTX in air	Ref.
	$T_{benzene50}$/°C	$T_{toluene50}$ /°C	$T_{o\text{-}xylene\ 30}$ /°C	(benzene/toluene or *o*-xylene)	
$SiO_{2(0.9)}Cu_{0.1}$	106	175	212	782 ppmv BTX in air	This study
$SiO_{2(0.95)}Cu_{0.05}$	124	191	225	782 ppmv BTX in air	This study
$SiO_{2(0.97)}Cu_{0.03}$	162	150	230	782 ppmv BTX in air	This study
5% Cu/γ-Al_2O_3	~310	~280	~275	800 ppmv BTX in air	**[17]**
15% Cu/γ-Al_2O_3	-	> 280	-	1000 ppmv BTX in air	**[17]**
5%Cu/TiO_2(rutile)	-	> 280	-	900 ppmv BTX in air	**[17]**
1.5% Au/ZnO	130	~235	~ 250	603 ppmv BTX, 10% O_2, 90% N_2	**[14]**
0.30%Pt/AC	131	-	-	640 ppmv BTX in air	**[8]**
3%Au/V2O5/TiO_2	~ 150	-	-	1206 ppmv BTX in air	**[22]**

According to the results shown in Figure 4, it was observed that the activity of the Cu/SiO_2 catalyst decreases in the order benzene > toluene> *o*-xylene. The conversion of benzene is higher than those of toluene and *o*-xylene same concentrations and reaction conditions.

This indicates that the catalytic activity in relation to aromatic compounds is highly dependent on the relative adsorption strength of the model compounds, the strength of the weakest carbon-hydrogen bond in the structure and the ionization potential of the methyl derivatives. It is well known that the oxidation of BTX compounds promoted by solid oxide catalysts can be proceed by the Mars-van Krevelen mechanism [8, 13, 14].

The $SiO_{2(0.9)}Cu_{0.1}$ exhibits higher activity compared with catalysts reported in the literature (5% and 15% Cu/γ-Al_2O_3, 5% Cu/TiO_2(rutile), 1.5% Au/ZnO and 3% Au/V_2O_5/TiO_2) with regard to BTX oxidation. The benzene conversion for the $SiO_{2(0.9)}Cu_{0.1}$ catalyst exceeded 85% at 150°C, which is comparable to the values for of 1.5% Au/ZnO and 3% Au/V_2O_5/TiO_2. The high catalytic activity could be attributed to the high stability observed due to a-greater energy of interaction between the Cu-SiO_2 catalysts in this study. Thus, the catalysts of the base Cu/SiO_2 proved to be highly efficient in the conversion of BTX compounds [8, 14, 17].

Figure 5 shows the conversion of BTX as a function of the reaction temperature for the catalyst studied. According to the results shown in Figure 5, it was observed that the activity of the WO_3 catalyst increases in the order *m*-xylene<*p*-xylene<toluene<benzene.

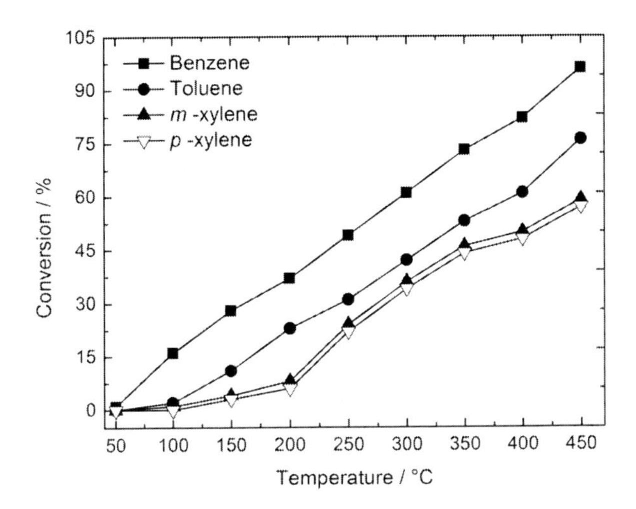

Figure 5. BTX conversion as a function of reaction temperature.

The results obtained in the catalyst characterization suggest that the high catalytic activity observed for the WO_3 catalyst can be attributed to the presence of the monoclinic WO_3, as verified by the infrared and X-ray analysis. Another possible contributing factor is a strong metal interaction, which could originate from the small lattice parameter difference between the (111), (020) and (002) lattice planes. The XPS analysis revealed the presence of three species of tungsten oxide with different oxidation states. The W^{+4}, W^{+5} and W^{+6} species present on the surface of the catalyst could react with active oxygen species favoring the oxidation process. Thus, the tungsten catalyst may increase the amount of oxygen vacancies and, in turn, the oxygen mobility and the catalytic activity. These vacancies are filled by the oxygen atoms that diffuse from the bulk to the surface of the catalyst, suggesting that the presence of crystalline defects enhances the mobility of the oxygen species. An increase in the amount of oxygen vacancies can enhance of the bulk and surface oxygen mobility, which appears to play an important role in oxidation reactions [4, 17, 23, 27].

The higher oxygen mobility facilitates the oxygen species migration across the catalyst structure, resulting in greater oxidation activity. It has been shown that the reduction in tungsten oxide ($W^{+6}/W^{+5}/W^{+4}$) is not due to a direct release of oxygen into the gas phase, but rather to the interaction which occurs between the surface of the catalyst and the hydrocarbon. These reactions are driven by an increased capacity for the spontaneous release of oxygen from the WO_3 system, even in the absence of a reducing agent. In the presence of the WO_3 species, the catalyst promotes the reduction of $W^{+6}/W^{+5}/W^{+4}$, this factor being favorable for the oxidation of hydrocarbons [2, 4, 11].

The results obtained suggest that the catalytic activity of aromatic compounds is dependent on several factors including: the strength of the carbon-hydrogen bond in the structure; the relative strength of the adsorption of each compound in question; and the ionization potential of these compounds as well as the methyl derivatives (benzene (9.24 eV), toluene (8.82 eV), *meta*-xylene and *ortho*-xylene (8.56 eV) and *para*-xylene (8.44 eV)). The dependence on the ionization potential of the compounds is more evident when comparing the results obtained for *m*-xylene and *p*-xylene. In the case of *m*-xylene there was a slight decrease in the degree of conversion and this can be explained by the lower ionization potential of the compound.

On comparing the catalytic activity of the WO_3 catalyst with that of the Cu/SiO_2catalysts described in the literature which are highly active in BTX oxidation, it can be seen that temperature $T_{benzene50}$ (benzene conversion up to 50%) is 250°C for WO_3 and 162°C for $SiO_{2(0.97)}Cu_{0.03}$ (Figure IV). In previously published studies we use a catalyst of high surface area, which may have favored the catalytic activity, while in this study we used a catalyst with a low surface area, suggesting that the catalytic activity is favored by the its oxygen storage capacity on the surface, this being very favorable for catalytic oxidation reactions. Tungsten oxide has a high oxygen storage capacity favoring catalytic oxidation reactions [4, 30, 31].

The WO_3 catalyst exhibited high catalytic activity in the oxidation of BTX (benzene, toluene, and xylenes) compounds. The benzene conversion in the presence of the WO_3 catalyst exceeded 70% and the toluene conversion exceeded 50% at 350°C. The results obtained for the studies with *meta* and *para* xylene suggest that the conversion of volatile aromatic compounds is also dependent on the ionization potential. The performance of the WO_3 catalyst may be due to a combination of several factors, including number of exposed active sites of tungsten oxide and greater oxygen mobility. Thus, it was verified that the catalyst based on WO_3 described herein is efficient in the conversion of BTX compounds.

References

[1] Rodrigues, T.S.D.S., Anderson G. M. ; Gonçalves, Mariana C. ; Fajardo, Humberto V. ; Balzer, Rosana ; Probst, Luiz F. D. ; Camargo,

Pedro H. C. (2015). AgPt Hollow Nanodendrites: Synthesis and Uniform Dispersion over SiO 2 Support for Catalytic Applications. *ChemNanoMat,* 1, 46-51.

[2] da Silva, A.G.M., Fajardo, H.V., Balzer, R., Probst, L.F.D., Prado, N.T., Camargo, P.H.C., Robles-Dutenhefner, P.A. (2016). Efficient ceria-silica catalysts for BTX oxidation: Probing the catalytic performance and oxygen storage. *Chem Eng J,* 286, 369-376.

[3] da Silva, A.G.M., Rodrigues, T.S., Taguchi, L.S.K., Fajardo, H.V., Balzer, R., Probst, L.F.D., Camargo, P.H.C. (2016). Pd-based nanoflowers catalysts: controlling size, composition, and structures for the 4-nitrophenol reduction and BTX oxidation reactions. *J Mater Sci,* 51 (1), 603-614.

[4] Balzer, R., Drago, V., Schreiner, W.H., Probst, L.F.D. (2014). Synthesis and Structure-Activity Relationship of a WO3 Catalyst for the Total Oxidation of BTX. *J Brazil Chem Soc,* 25 (11), 2026-2031.

[5] Balzer, R., Probst, L.F.D., Drago, V., Schreiner, W.H., Fajardo, H.V. (2014). Catalytic Oxidation of Volatile Organic Compounds (N-Hexane, Benzene, Toluene, O-Xylene) Promoted By Cobalt Catalysts Supported On gamma-Al2O3-CeO2. *Braz J Chem Eng,* 31 (3), 757-769.

[6] Araujo, V.D., de Lima, M.M., Cantarero, A., Bernardi, M.I.B., Bellido, J.D.A., Assaf, E.M., Balzer, R., Probst, L.F.D., Fajardo, H.V. (2013). Catalytic oxidation of n-hexane promoted by Ce(1-x)Cu(x)O2 catalysts prepared by one-step polymeric precursor method. *Mater Chem Phys,* 142 (2-3), 677-681.

[7] Balzer, R., Drago, V., Schreiner, W.H., Probst, L.F.D. (2013). Removal of BTX Compounds in Air by Total Catalytic Oxidation Promoted by Catalysts Based on SiO2(1-x)Cux. *J Brazil Chem Soc,* 24 (10), 1592-1598.

[8] Wu, J.C.S., Lin, Z.A., Tsai, F.M., Pan, J.W. (2000). Low-temperature complete oxidation of BTX on Pt/activated carbon catalysts. *Catal Today,* 63 (2-4), 419-426.

[9] Wu, J.C.S., Sung, H.C., Lin, Y.F., Lin, S.L. (2000). Removal of tar base from coal tar aromatics employing solid acid adsorbents. *Sep Purif Technol,* 21 (1-2), 145-153.

[10] Balzer, R., Probst, L.F.D., Cantarero, A., de Lima, M.M., Araujo, V.D., Bernardi, M.I.B., Avansi, W., Arenal, R., Fajardo, H.V. (2015). Ce1-xCoxO2 Nanorods Prepared by Microwave-Assisted Hydrothermal Method: Novel Catalysts for Removal of Volatile Organic Compounds. *Sci Adv Mater,* 7 (7), 1406-1414.

[11] da Silva, A.G.M., Fajardo, H.V., Balzer, R., Probst, L.F.D., Lovon, A.S.P., Lovon-Quintana, J.J., Valenca, G.P., Schreine, W.H., Robles-Dutenhefner, P.A. (2015). Versatile and efficient catalysts for energy and environmental processes: Mesoporous silica containing Au, Pd and Au-Pd. *J Power Sources,* 285, 460-468.

[12] Pozzobom, I.E.F., de Moraes, G.G., Balzer, R., Probst, L.F.D., Triches, E.S., de Oliveira, A.P.N. (2015). Glass-Ceramics Foam for Hydrogen Production. *Chem Engineer Trans,* 43, 1789-1794.

[13] Wu, H.J., Wang, L.D. (2011). Shape effect of microstructured CeO2 with various morphologies on CO catalytic oxidation. *Catal Commun,* 12 (14), 1374-1379.

[14] Wu, H.J., Wang, L.D., Zhang, J.Q., Shen, Z.Y., Zhao, J.H. (2011). Catalytic oxidation of benzene, toluene and p-xylene over colloidal gold supported on zinc oxide catalyst. *Catal Commun,* 12 (10), 859-865.

[15] Dhada, I., Nagar, P.K., Sharma, M. (2016). Photo-catalytic oxidation of individual and mixture of benzene, toluene and p-xylene. *Int J Environ Sci Te,* 13 (1), 39-46.

[16] Wang, W.Z., Wang, H.L., Zhu, T.L., Fan, X. (2015). Removal of gas phase low-concentration toluene over Mn, Ag and Ce modified HZSM-5 catalysts by periodical operation of adsorption and non-thermal plasma regeneration. *J Hazard Mater,* 292, 70-78.

[17] Kim, S.C. (2002). The catalytic oxidation of aromatic hydrocarbons over supported metal oxide. *J Hazard Mater,* 91 (1-3), 285-299.

[18] Arabatzis, I.M., Stergiopoulos, T., Andreeva, D., Kitova, S., Neophytides, S.G., Falaras, P. (2003). Characterization and photocatalytic activity of Au/$TiO_{(2)}$ thin films for azo-dye degradation. *J Catal,* 220 (1), 127-135.

[19] Morales-Torres, S., Carrasco-Marin, F., Perez-Cadenas, A.F., Maldonado-Hodar, F.J. (2015). Coupling Noble Metals and Carbon Supports in the Development of Combustion Catalysts for the Abatement of BTX Compounds in Air Streams. *Catalysts,* 5 (2), 774-799.

[20] Sinha, S., Raj, A., Al Shoaibi, A.S., Chung, S.H. (2015). Reaction Mechanism for m-Xylene Oxidation in the Claus Process by Sulfur Dioxide. *J Phys Chem A,* 119 (38), 9889-9900.

[21] Raj, A., Sinha, S. (2015). Reaction mechanism for the oxidation of aromatic contaminants present in feed gas to Claus process. *Enrgy Proced,* 66, 61-64.

[22] Andreeva, D., Nedyalkova, R., Ilieva, L., Abrashev, M.V. (2003). Nanosize gold-ceria catalysts promoted by vanadia for complete benzene oxidation. *Appl Catal a-Gen,* 246 (1), 29-38.

[23] Idakiev, V., Ilieva, L., Andreeva, D., Blin, J.L., Gigot, L., Su, B.L. (2003). Complete benzene oxidation over gold-vanadia catalysts supported on nanostructured mesoporous titania and zirconia. *Appl Catal a-Gen,* 243 (1), 25-39.

[24] Solsona, B., Davies, T.E., Garcia, T., Vazquez, I., Dejoz, A., Taylor, S.H. (2008). Total oxidation of propane using nanocrystalline cobalt oxide and supported cobalt oxide catalysts. *Appl Catal B-Environ,* 84 (1-2), 176-184.

[25] Zhao, Z.K., Jin, R.H., Bao, T., Lin, X.L., Wang, G.R. (2011). Mesoporous ceria-zirconia supported cobalt oxide catalysts for CO preferential oxidation reaction in excess H-2. *Appl Catal B-Environ,* 110, 154-163.

[26] Warang, T., Patel, N., Santini, A., Bazzanella, N., Kale, A., Miotello, A. (2012). Pulsed laser deposition of Co3O4 nanoparticles assembled coating: Role of substrate temperature to tailor disordered to crystalline phase and related photocatalytic activity in degradation of methylene blue. *Appl Catal a-Gen,* 423, 21-27.

[27] Kang, M., Song, M.W., Lee, C.H. (2003). Catalytic carbon monoxide oxidation over CoOx/CeO2 composite catalysts. *Appl Catal a-Gen,* 251 (1), 143-156.

[28] Luo, J.Y., Meng, M., Li, X., Li, X.G., Zha, Y.Q., Hu, T.D., Xie, Y.N., Zhang, J. (2008). Mesoporous Co(3)O(4)-CeO(2) and Pd/Co(3)O(4)-CeO(2) catalysts: Synthesis, characterization and mechanistic study of their catalytic properties for low-temperature CO oxidation. *J Catal,* 254 (2), 310-324.

[29] Todorova, S., Kadinov, G., Tenchev, K., Caballero, A., Holgado, J.P., Pereniguez, R. (2009). Co3O4 + CeO2/SiO2 Catalysts for n-Hexane and CO Oxidation. *Catal Lett,* 129 (1-2), 149-155.

[30] Anisia, K.S., Kumar, A. (2007). Oxidation of cyclohexane with molecular oxygen in presence of characterized macrocyclic heteronuclear FeCu complex catalyst ionically bonded to zirconium pillared montmorillonite clay. *J Mol Catal a-Chem,* 271 (1-2), 164-179.

[31] Aouissi, A. (2010). Transformation of n-Heptane by Bronsted Acidic Sites Over 12-Tungstosilicic Acid. *Asian J Chem,* 22 (6), 4924-4930.

RELATED NOVA PUBLICATIONS

VOLATILE ORGANIC COMPOUNDS

Jonathan C. Hanks and Sara O. Louglin

ISBN: 978-1-61324-156-1
Publication Date: 2011

Volatile organic compounds (VOCs) refers to organic chemical compounds which have significant vapor pressures. This new book presents current research in volatile organic compounds, including the effect of transition metals on the reductive dechlorination of COCs by iron-bearing soil minerals; photocatalytic degradtion of VOC gases using a short wavelength UV light and water droplets; catalytic incineration of VOCs; transport of VOCs in polymers; sources and elimination of VOCs and in-vivo analysis of palm wine VOCs by proton transfer reaction-mass spectrometry.

VOLATILE ORGANIC COMPOUNDS: EMISSION, POLLUTION AND CONTROL

Khaled Chetehouna (Bourges Higher School of Engineering, Bourges, France)
ISBN: 978-1-63117-862-7
Publication Date: 2014

Volatile Organic Compounds (VOCs) have anthropogenic and biogenic origins. At the Earth's scale, the natural sources represent a great part of the total VOCs present in the atmosphere but in industrialized regions, anthropogenic ones become the majority due to the various human activities related mainly to chemical industries (liquid fuels, solvents, thinners, detergents, degreasers, cleaners and lubricants). Almost all VOCs have effects on human health and many of them are even carcinogenic. It is also known that the VOCs can affect the central nervous system and may have mutagenic effects. Apart from human health, they also play an important role towards the environment, especially in the atmospheric pollution processes. Indeed, VOCs emissions lead to the promotion of photochemical reactions in the atmosphere (ozone formation, depletion of the stratospheric ozone layer and formation of photochemical smog). The present book gathers and presents some current research from across the world conducted by scientific experts in their fields. In seven valuable contributions, it deals with the emission and the environmental impact as well as the control of the Volatile Organic Compounds.

INDEX

#

A

B

C

D

E

F

M

N

O

P

Q

R

S

T

U

V

W

X